Advances in Skeletal Muscle Health and Disease

Edited by Fabio Arturo Iannotti

Published in London, United Kingdom

Advances in Skeletal Muscle Health and Disease
http://dx.doi.org/10.5772/intechopen.100953
Edited by Fabio Arturo Iannotti

Contributors
Ridvan İşik, Ferhat Ege, Mark Lambrechts, Radenka Kuzmanić Šamija, Marta Plejić, Ashwin Parchani, Monika Pathania, Shiana Singh, Aditya Sudan, Arpit Singh, Fabio Arturo Iannotti

First published in London, United Kingdom, 2023 by IntechOpen
IntechOpen is the global imprint of INTECHOPEN LIMITED, registered in England and Wales, registration number: 11086078, 5 Princes Gate Court, London, SW7 2QJ, United Kingdom
Printed in Croatia

British Library Cataloguing-in-Publication Data
A catalogue record for this book is available from the British Library

Additional hard and PDF copies can be obtained from orders@intechopen.com

Advances in Skeletal Muscle Health and Disease
Edited by Fabio Arturo Iannotti
p. cm.

This title is part of the Physiology Book Series, Volume 18
Topic: Human Physiology
Series Editor: Tomasz Brzozowski
Topic Editor: Kunihiro Sakuma
Associate Topic Editor: Kotomi Sakai

Print ISBN 978-1-80356-272-8
Online ISBN 978-1-80356-273-5
eBook (PDF) ISBN 978-1-80356-274-2
ISSN 2631-8261

IntechOpen Book Series
Physiology
Volume 18

Aims and Scope of the Series

Modern physiology requires a comprehensive understanding of the integration of tissues and organs throughout the mammalian body, including the cooperation between structure and function at the cellular and molecular levels governed by gene and protein expression. While a daunting task, learning is facilitated by identifying common and effective signaling pathways mediated by a variety of factors employed by nature to preserve and sustain homeostatic life. As a leading example, the cellular interaction between intracellular concentration of Ca+2 increases, and changes in plasma membrane potential is integral for coordinating blood flow, governing the exocytosis of neurotransmitters, and modulating gene expression and cell effector secretory functions. Furthermore, in this manner, understanding the systemic interaction between the cardiovascular and nervous systems has become more important than ever as human populations' life prolongation, aging and mechanisms of cellular oxidative signaling are utilised for sustaining life. Altogether, physiological research enables our identification of distinct and precise points of transition from health to the development of multimorbidity throughout the inevitable aging disorders (e.g., diabetes, hypertension, chronic kidney disease, heart failure, peptic ulcer, inflammatory bowel disease, age-related macular degeneration, cancer). With consideration of all organ systems (e.g., brain, heart, lung, gut, skeletal and smooth muscle, liver, pancreas, kidney, eye) and the interactions thereof, this Physiology Series will address the goals of resolving (1) Aging physiology and chronic disease progression (2) Examination of key cellular pathways as they relate to calcium, oxidative stress, and electrical signaling, and (3) how changes in plasma membrane produced by lipid peroxidation products can affect aging physiology, covering new research in the area of cell, human, plant and animal physiology.

Meet the Series Editor

Prof. Dr. Thomas Brzozowski works as a professor of Human Physiology and is currently a Chairman at the Department of Physiology and is V-Dean of the Medical Faculty at Jagiellonian University Medical College, Cracow, Poland. His primary area of interest is physiology and pathophysiology of the gastrointestinal (GI) tract, with a major focus on the mechanism of GI mucosal defense, protection, and ulcer healing. He was a postdoctoral NIH fellow at the University of California and the Gastroenterology VA Medical Center, Irvine, Long Beach, CA, USA, and at the Gastroenterology Clinics Erlangen-Nuremberg and Munster in Germany. He has published 290 original articles in some of the most prestigious scientific journals and seven book chapters on the pathophysiology of the GI tract, gastroprotection, ulcer healing, drug therapy of peptic ulcers, hormonal regulation of the gut, and inflammatory bowel disease.

Meet the Volume Editor

Dr. Fabio Arturo Iannotti received his master's degree in Medical Biotechnology at the University Federico II, Italy in 2006. In 2010, Dr. Iannotti graduated with a Ph.D. in Neuroscience. During this time, he discovered the pro-differentiation and protective role of voltage-gated Kv7 K+ channels in skeletal muscle cells. Part of this research was carried out at the University of California-Davis, USA (2009–2010). In 2011, he began his postdoc at the Institute of Biomolecular Chemistry (ICB)/(CNR), initiating pioneering studies on the role of the endocannabinoid system as well as plant-derived cannabinoids in skeletal muscle disorders. In 2012–2013, he was invited as visiting researcher at the University of Reading, UK. In 2014, he became a permanent researcher at the ICB.

Contents

Preface

The skeletal muscle system includes more than 600 muscles and accounts for approximately 40% of a person's total body weight. Skeletal muscles allow the body to maintain posture and stability, move, breathe, eat, and regulate glucose and lipid metabolism. Furthermore, skeletal muscles are the primary storage site for amino acids (AAs), which are released "on demand" and used as fuel by other organs and tissues of the body. Consequently, diseases that affect the skeletal muscles significantly impact general health and well-being.

Myopathy is a general term that refers to a large and heterogeneous group of rare diseases affecting skeletal muscles leading to various patterns of weakness and dysfunction. These conditions can be roughly classified into two main categories: inherited and acquired myopathies. Inherited myopathies include several forms of dystrophic (e.g., myotonic dystrophy, Duchenne and Becker muscular dystrophy, congenital and facioscapulohumeral muscular dystrophy) and non-dystrophic (e.g., mitochondrial, congenital, and metabolic) myopathies, which are caused by mutations (most frequently deletions, insertions, and missense mutations) in genes encoding for proteins essential for muscle structure and function with X-linked, autosomal-recessive, or autosomal-dominant inheritance patterns. Acquired myopathies encompass endocrine, inflammatory, and infectious myopathies as well as those that can be caused by prolonged use of certain drugs (i.e., statins) or alcohol abuse.

Unfortunately, for most myopathies, no satisfactory method of diagnosis, prevention, or treatment exists, although experimental therapies have made astonishing advances over the years. The mechanisms of repair of skeletal muscle injury have been a long-term concern in basic, clinical, and translational research. Extensive research has been done at cellular and molecular levels and in animal models of myopathies, and we know a lot more about the processes driving muscle degeneration. However, many aspects of myopathies have not been completely solved or remain fully elusive.

That said, novel insights into the mechanisms underlying the pathogenesis of skeletal muscle disorders are continuously emerging and therefore research in this field has been moved forward abruptly from a belief that the severity and progression of skeletal muscle diseases besides the absence or defective function of specific structural proteins is also driven by alteration of signalling molecules, epigenetic and post-translational modifications, organelles dysfunction, and interaction with other tissues and organs of the body. This exciting new paradigm is important to shed light on those pathological aspects of myopathies that remain elusive.

This book collects and disseminates knowledge about the recent discoveries in skeletal muscle complex pathophysiology, development in molecular diagnostics, and promising progress in treatments for skeletal disease. It was my honour to work with colleagues from around the world to publish this volume.

To my special wife Elisa and my
beloved daughters Chiara and Francesca.
My love, my life, my light.

Fabio Arturo Iannotti
Institute of Biomolecular Chemistry (ICB),
Research National Council (CNR),
Pozzuoli (NA), Italy

Section 1

Introduction

Chapter 1

Introductory Chapter: Skeletal Muscle Disorders – Emerging New Avenues for the Diagnosis and Treatment

Fabio Arturo Iannotti

1. Introduction

After years of intense clinical and experimental research using structural, biological, and biochemical experimental procedures, it is clear that the etiology and severity of skeletal muscle disorders, which encompass a large and complex heterogeneous group of diseases known as myopathies, are determined by a complex relationship between genetic and environmental factors and not last by multiple organ dysfunctions. Myopathies generally are grouped into two main categories: acquired and genetically determined (hereditary). All of them have a rare or extremely rare frequency. This often makes difficult the final diagnosis and development of a treatment leading to an inadequate response to the needs of patients and their families.

2. Current and promising therapeutic options for muscular dystrophies

Included in the acquired group of muscle disorders are endocrine, metabolic, inflammatory (polymyositis, dermatomyositis, and inclusion body myositis) and toxic myopathies. While hereditary myopathies include congenital, metabolic, mitochondrial, myotonias, and muscular dystrophies that are caused by mutations in different genes-encoding proteins that play important roles in muscle structure and function with X-linked, autosomal-recessive, or autosomal-dominant inheritance patterns. Among them, Duchenne (DMD) and Becker (BMD) muscular dystrophies are the most frequent whose prevalence is estimated at 4.8 per 100,000 people (95 CI 3.6–6.3 per 100,000 people) and 1.6 per 100,000 people (95 CI 1.1–2.4 per 100,000 people), respectively [1]. Other forms of inherited myopathies (extremely rare) include sarcoglicanopathies, and congenital and metabolic myopathies. The progressive irreversible loss of muscle fibers resulting from repetitive cycles of degeneration, necrosis, regeneration, and eventually fibrosis and fat is a common pathological event of muscular dystrophies.

Unfortunately, a cure for DMD as well as other muscular dystrophies is not available, although innovative experimental therapies such as gene therapy, CRISPR/Cas9, exon-skipping, cell therapy have made astonishing advances over the years [2]. As a consequence, the use of anti-inflammatory drugs mostly including steroids remains the mainstay of palliative care. In particular, prednisolone (PRED) and deflazacort (DFZ)

IntechOpen

were shown to produce the beneficial effects on the preservation of functional abilities in several mouse models of DMD and randomized controlled trials [3, 4]. However, the long-term daily use of corticosteroids may cause in young boys metabolic and growth-related side effects, pubertal delay, and increased risk of vertebral fractures [5, 6]. But why do novel therapies achieve limited success? Undoubtedly, introducing changes into specific cells of the body by targeting a specific gene or part of it is feasible but often challenging, especially in skeletal muscle cells/tissues that are extremely vast accounting in fact for almost half of the human body mass. Moreover, the success of experimental therapies is limited by the complex structure of skeletal muscles organized in numerous bundles of muscle fiber (myofibers) containing millions of myofibrils. Not least, the type of muscles affected by the disease may vary with the type of hereditary disorder [7]. Besides skeletal muscles, cardiac problems in hereditary muscle diseases are frequent and include cardiomyopathies, defects of cardiac conductions with or without primary myocardial muscle involvement, and arrhythmias [8]. Remarkably, there is also evidence of central nervous system involvement in skeletal muscle disorders [9]. In this respect, progresses in neuroimaging and electrophysiological techniques demonstrated that the absence or reduced expression of dystrophin leads to neurological, cognitive, and neuroanatomical alterations to varying degrees of severity in a significant number of DMD patients [10]. However, the current knowledge about the molecular mechanisms underlying behavioral and cognitive deficits in DMD patients remains very modest [11]. The most accredited hypothesis is that the loss of dystrophin causes a displacement of key proteins at the synaptic level with consequent disruptions of their functions [10]. Consequently, the production of pro-inflammatory cytokines contributes to memory defects and neurochemical alterations [12]. In both animal models and patients with DMD, an increase in pro-inflammatory factors including interleukin 6 (IL-6), interleukin 1β (IL-1β), and tumor necrosis factor-α (TNF-α) was found in muscle tissues as well as blood samples [13, 14]. Furthermore, Nico and colleagues showed that mdx mice, a validated animal model of DMD, present an increase in blood-brain barrier permeability associated with an increased matrix-metalloproteinase-2 and matrix-metalloproteinase-9 expression. These changes may indeed facilitate the process of neuroinflammation in DMD patients, besides the absence of dystrophin in brain tissue [15, 16]. Therefore, investigations on pathological mechanisms underlying skeletal muscle diseases followed by targeted clinical interventions are urgently needed.

In the last decades, important discoveries have emerged. This year Guglieri and colleagues published a study on JAMA after carrying out a randomized clinical trial that included 196 boys with DMD receiving 3 corticosteroid regimens (0.75 mg/kg of daily prednisone, 0.90 mg/kg of daily deflazacort, or 0.75 mg/kg of intermittent prednisone for 10 days on and then 10 days off). Remarkably, the results of this study demonstrated that daily prednisone or deflazacort resulted in significantly better outcomes compared with intermittent prednisone; there was no significant difference between the two daily regimens [17]. This finding represents an outstanding contribution to maximizing the use and beneficial outcomes of glucocorticoids not only in DMD but also in other myopathies. Previously existing data instead revealed that an intermittent regimen of oral prednisolone for two consecutive days per week in mdx mice resulted in an increased strength over time and improved survival between 80 and 104 weeks of age. Intermittent injection of prednisone or deflazacort at a minimal dose of once-weekly comparably benefitted sarcolemmal repair, fibrosis, and immune infiltrations as daily steroids in short-term experiments, while once-weekly versus daily prednisone induced opposite epigenetic and metabolic changes [18].

In an attempt to fight skeletal muscle diseases, several disease-modifying agents have been recently identified at the cellular level and are being investigated to offer the chance of a cure. For example, a growing number of studies show that dysregulation of autophagy and mitophagy aggravates muscle damage and contributes to disease progression in DMD. Accordingly, normalization of defective autophagy/mitophagy was suggested as a novel strategy to reduce muscle damage and promote muscle regeneration [19, 20]. Remarkably, Moore and colleagues demonstrated aberrant mitochondrial morphology, reduced number of mitochondrial cristae, and large mitochondrial vacuoles from both male and female mdx mice before the onset of muscle damage, thus reinforcing the evidence that these targets may yield novel therapeutic targets for the prevention and management of DMD [21]. It is also known that autophagy and inflammation exist in a complex relationship. Autophagy plays a critical role in the development, homeostasis, and survival of inflammatory cells, including macrophages, neutrophils, and lymphocytes, which play critical roles in the development and pathogenesis of inflammation. Inflammatory factors (e.g., NF-κB), are in turn critical regulator of autophagy in skeletal muscle [22].

Yet, fascinating studies revealed that fibroadipogenic progenitors (FAP) influence the regeneration potential of satellite cells during disease progression in mdx mice and mediate HDACi ability to selectively promote regeneration at the early stages of the disease [23, 24].

Again, in the last years, a lot has been learned about the association between gut microbiota with skeletal muscle homeostasis and function [25, 26]. However, the mechanisms through which intestinal microorganisms exert their influence on skeletal muscle remain largely undefined. To date, nothing is known about the potential implication of gut microbiota in skeletal muscle disorders. In this context, my research group has recently obtained evidence that fecal microbiota along with circulating levels of their metabolites is significantly altered between mdx and healthy mice. Therefore, we demonstrated that gut microbiota in dystrophic mice represent a novel disease-modifying agent to target for the development of new treatments.

In conclusion, this book topic is devoted to encompassing preclinical research and clinical studies on hereditary and acquired skeletal muscle disorders, ranging from molecular mechanisms to implications of clinical practice.

Author details

Fabio Arturo Iannotti
Institute of Biomolecular Chemistry (ICB)/Consiglio Nazionale delle Ricerche (CNR), Pozzuoli (Na), Italy

*Address all correspondence to: fabio.iannotti@icb.cnr.it

References

[1] Salari N, Fatahi B, Valipour E, Kazeminia M, Fatahian R, Kiaei A, et al. Global prevalence of Duchenne and Becker muscular dystrophy: A systematic review and meta-analysis. Journal of Orthopaedic Surgery and Research. 2022;**17**:96. DOI: 10.1186/s13018-022-02996-8

[2] Sheikh O, Yokota T. Advances in genetic characterization and genotype-phenotype correlation of Duchenne and Becker muscular dystrophy in the personalized medicine era. Journal of Personalized Medicine. 2020;**10**:111. DOI: 10.3390/jpm10030111

[3] Bushby K, Finkel R, Birnkrant DJ, Case LE, Clemens PR, Cripe L, et al. Diagnosis and management of Duchenne muscular dystrophy, part 1: Diagnosis, and pharmacological and psychosocial management. The Lancet Neurology. 2010;**9**:77-93. DOI: 10.1016/S1474-4422(09)70271-6

[4] Griggs RC, Miller JP, Greenberg CR, Fehlings DL, Pestronk A, Mendell JR, et al. Efficacy and safety of deflazacort vs prednisone and placebo for Duchenne muscular dystrophy. Neurology. 2016;**87**:2123-2131. DOI: 10.1212/WNL.0000000000003217

[5] Lamb MM, West NA, Ouyang L, Yang M, Weitzenkamp D, James K, et al. Corticosteroid treatment and growth patterns in ambulatory males with Duchenne muscular dystrophy. The Journal of Pediatrics. 2016;**173**:207-213.e3. DOI: 10.1016/j.jpeds.2016.02.067

[6] Bello L, Gordish-Dressman H, Morgenroth LP, Henricson EK, Duong T, Hoffman EP, et al. CINRG investigators prednisone/prednisolone and deflazacort regimens in the CINRG Duchenne natural history study. Neurology. 2015;**85**:1048-1055. DOI: 10.1212/WNL.0000000000001950

[7] Chu X, Du K, Tang Y, Zhao X, Yu M, Zheng Y, et al. Skeletal muscle involvement pattern of hereditary transthyretin amyloidosis: A study based on muscle MRI. Frontiers in Neurology. 2022;**13**:851190

[8] Giuliani L, Di Toro A, Urtis M, Smirnova A, Concardi M, Favalli V, et al. Hereditary muscle diseases and the heart: The Cardiologist's perspective. European Heart Journal Supplements. 2020;**22**:E13-E19. DOI: 10.1093/eurheartj/suaa051

[9] Reimann J, Kornblum C. Towards central nervous system involvement in adults with hereditary myopathies. Journal of Neuromuscular Diseases. 2020;**7**:367-393. DOI: 10.3233/JND-200507

[10] Stefano MED, Ferretti V, Mozzetta C. Synaptic alterations as a neurodevelopmental trait of Duchenne muscular dystrophy. Neurobiology of Disease. 2022;**168**:105718. DOI: 10.1016/j.nbd.2022.105718

[11] Peristeri E, Aloizou A-M, Keramida P, Tsouris Z, Siokas V, Mentis A-FA, et al. Cognitive deficits in myopathies. International Journal of Molecular Sciences. 2020;**21**:3795. DOI: 10.3390/ijms21113795

[12] Donzis EJ, Tronson NC. Modulation of learning and memory by cytokines: Signaling mechanisms and long term consequences. Neurobiology of Learning and Memory. 2014;**0**:68-77. DOI: 10.1016/j.nlm.2014.08.008

[13] Comim CM, Ventura L, Freiberger V, Dias P, Bragagnolo D, Dutra ML, et al.

Neurocognitive impairment in Mdx Mice. Molecular Neurobiology. 2019;**56**:7608-7616. DOI: 10.1007/s12035-019-1573-7

[14] Wada E, Tanihata J, Iwamura A, TakedaS,HayashiYK,MatsudaR.Treatment with the anti-IL-6 receptor antibody attenuates muscular dystrophy via promoting skeletal muscle regeneration in dystrophin-/utrophin-deficient mice. Skeletal Muscle. 2017;7:23. DOI: 10.1186/s13395-017-0140-z

[15] Nico B, Corsi P, Ria R, Crivellato E, Vacca A, Roccaro AM, et al. Increased matrix-metalloproteinase-2 and matrix-metalloproteinase-9 expression in the brain of dystrophic mdx mouse. Neuroscience. 2006;**140**:835-848. DOI: 10.1016/j.neuroscience.2006.02.077

[16] Nico B, Roncali L, Mangieri D, Ribatti D. Blood-brain barrier alterations in MDX mouse, an animal model of the Duchenne muscular dystrophy. Current Neurovascular Research. 2005;**2**:47-54. DOI: 10.2174/1567202052773481

[17] Guglieri M, Bushby K, McDermott MP, Hart KA, Tawil R, Martens WB, et al. Effect of different corticosteroid dosing regimens on clinical outcomes in boys with Duchenne muscular dystrophy: A randomized clinical trial. JAMA. 2022;**327**:1456-1468. DOI: 10.1001/jama.2022.4315

[18] Quattrocelli M, Zelikovich AS, Salamone IM, Fischer JA, McNally EM. Mechanisms and clinical applications of glucocorticoid steroids in muscular dystrophy. Journal of Neuromuscular Diseases. 2021;**8**:39-52. DOI: 10.3233/JND-200556

[19] De Palma C, Perrotta C, Pellegrino P, Clementi E, Cervia D. Skeletal muscle homeostasis in Duchenne muscular dystrophy: Modulating autophagy as a promising therapeutic strategy. Frontiers in Aging Neuroscience. 2014;**6**:188. DOI: 10.3389/fnagi.2014.00188

[20] Sandri M, Coletto L, Grumati P, Bonaldo P. Misregulation of autophagy and protein degradation systems in myopathies and muscular dystrophies. Journal of Cell Science. 2013;**126**:5325-5333. DOI: 10.1242/jcs.114041

[21] Moore TM, Lin AJ, Strumwasser AR, Cory K, Whitney K, Ho T, et al. Mitochondrial dysfunction is an early consequence of partial or complete dystrophin loss in mdx mice. Frontiers in Physiology. 2020;**11**:690

[22] Xia Q, Huang X, Huang J, Zheng Y, March ME, Li J, et al. The role of autophagy in skeletal muscle diseases. Frontiers in Physiology. 2021;**12**:638983

[23] Mozzetta C, Consalvi S, Saccone V, Tierney M, Diamantini A, Mitchell KJ, et al. Fibroadipogenic progenitors mediate the ability of HDAC inhibitors to promote regeneration in dystrophic muscles of young, but not old mdx mice. EMBO Molecular Medicine. 2013;**5**:626-639. DOI: 10.1002/emmm.201202096

[24] Sandonà M, Consalvi S, Tucciarone L, De Bardi M, Scimeca M, Angelini DF, et al. HDAC inhibitors tune MiRNAs in extracellular vesicles of dystrophic muscle-resident mesenchymal cells. EMBO Reports. 2020;**21**:e50863. DOI: 10.15252/embr.202050863

[25] Bäckhed F, Manchester JK, Semenkovich CF, Gordon JI. Mechanisms underlying the resistance to diet-induced obesity in germ-free mice. Proceedings. National Academy of Sciences. United States of America. 2007;**104**:979-984. DOI: 10.1073/pnas.0605374104

[26] Lustgarten MS. The role of the gut microbiome on skeletal muscle mass and physical function. Update. Frontiers in Physiology. 2019;**2019**:10. DOI: 10.3389/fphys.2019.01435

Section 2

Myopathies: Skeletal Muscle Complications and Beyond

Chapter 2

Musculoskeletal Manifestations in Sjogren's Syndrome

Ridvan İşik and Ferhat Ege

Abstract

Sjögren's syndrome (SS) is a chronic, autoimmune, inflammatory disease characterized by lymphocytic infiltration, destruction and dysfunction of the exocrine glands. Sjögren's syndrome can be described as primary or secondary, depending on whether it occurs alone or in association with other systemic autoimmune diseases. Systemic manifestations of SS involve the musculoskeletal system. SS can be seen in association with both joint and muscle manifestations, including arthralgia and arthritis, as well as myopathy, which is usually asymptomatic. Besides, it may include bone metabolic disorders, fatigue and fibromyalgia. The diagnosis of Sjögren's syndrome is based on characteristic clinical signs and symptoms. The etiology and pathogenesis of SS is elusive and has not yet been clarified. There is no curative treatment for SS, thus the aim in the treatment of SS is to alleviate the symptoms.

Keywords: musculoskeletal, joints, fibromyalgia, fatigue, bones

1. Introduction

Sjögren's syndrome (SS) is a chronic, autoimmune, systemic, inflammatory disease that develops as a result of lymphocytic infiltration, destruction and dysfunction of the exocrine glands, and the lacrimal and salivary glands in particular. In 1933, Swedish ophthalmologist Henrik Sjögren described the clinical and histological findings associated with SS in 19 patients with rheumatoid arthritis, 13 of whom had dry mouth and dry eye symptoms [1]. SS predominantly affects middle-aged women who are within the fourth to sixth decade of their lives. The male/female ratio in patients with SS is approximately 6:1 to 9:1. SS is usually diagnosed in the fifth decade of life, but the first symptoms may appear years before diagnosis [2]. The incidence and prevalence rates of SS were estimated approximately as 6.92 cases per 100.000 persons/year and as 60.82 cases per 100.000 persons, respectively [3]. Geographical location and ethnicity have a strong influence on the biological and clinical phenotype of the disease. The onset of diagnosis of Sjogren syndrome and the gender preference may be affected by racial variation: the diagnosis can be accomplished up to 7 years earlier in patients of black/African American origins compared to Caucasians. Furthermore, the female to male ratio might reach 27:1 in patients of Asian descent [4]. There are two types of SS; first type, that is the primary SS, is the type without any concomitant connective tissue disorder, whereas the second type, that is the secondary SS, is the type observed together with other autoimmune diseases such as systemic lupus erythematosus (SLE), rheumatoid arthritis (RA) or systemic sclerosis (SSc). SS is characterized by a wide spectrum of signs and symptoms, ranging from glandular involvement, structural symptoms, extraglandular manifestations and systemic autoimmune features. SS can be seen in association with both joint and muscle manifestations, including arthralgia and arthritis, as well as myopathy, which is usually asymptomatic. SS mostly involves the bone, the synovium and the cartilage tissue [5].

2. Pathogenesis

The etiology and pathogenesis of SS has not been clarified, as is the case with other autoimmune diseases [6]. Until now, it has been widely accepted that environmental factors play a role in the pathogenesis of SS [7]. Nevertheless, a thorough review of the recently published studies available in the literature revealed that the complex interaction between epithelial cells and targets of the autoimmune response and genetic and epigenetic changes also play a role in the pathogenesis of SS, in addition to the activated innate and adaptive immune system [8, 9].

The genetic component of SS are yet to shed light on. However, recent studies have begun to elucidate the familial links of the disease, identify specific risk alleles, and even classify patients according to their global gene expression levels. In these studies, many risk alleles for SS have been highlighted. Identification of these risk alleles helps in early diagnosis and choice of treatment options. Patients with extra-glandular manifestations (EGM) were found to have higher expression of genes involved in innate (apoptosis, TLR and interferon signaling) and adaptive (T and B cell activation) immune responses that play a key role in SS. On the other hand, patients with glandular features (GF) and diffuse pain (WP) were found to have the highest differential gene expression related to sensory perception and pain [10]. In complex diseases such as SS, the on and off signals of gene expression related to inflammatory pathways are managed by epigenetic mechanisms. Several epigenetic mechanisms, i.e., DNA methylation, miRNAs, and lncRNAs, contribute to turning on and off the expression of genes involved in inflammatory pathways and may target amelioration of SS therapy [11].

Considerable efforts have been made to elucidate the role of the innate immune system in the pathogenesis of SS. Plasmacytoid dendritic cells (pDC) are the predominant type I interferon (IFN) producing cells. The transcriptional profile of SS plasmacytoid dendritic cells (pDCs) has been investigated and interestingly, it was found to be associated with enhanced cytokine production of pDCs. TLR7 dominant innate immunity may be associated with the development of sialadenitis in SS. Additionally, the few evidence supporting the role of TLR7-dominant innate immunity in the development of sialadenitis in SS [12].

The role of B and T cell subsets, particularly, of T follicular helper cells (Tfh) and of their regular counterparts, that is the T follicular regulatory cells (Tfr) cells, has been extensively investigated. The recent data on the subject revealed an increase in the proportion of Tfr and Tfh cells in SS patients as compared to the healthy control subjects, and an imbalance between proinflammatory and immune regulatory pathways in SS. Tertiary or ectopic lymphoid structures (TLS) are lymphoid clusters of T and B cells that form in non-lymphoid organs in response to chronic inflammation. TLS form in the target organ of autoimmune diseases, including SS, and is generally associated with worse disease progression. New insights into TLS formation and care are paving the way for new therapeutic approaches to SS [13].

3. Articular involvement

Most SS patients exhibit musculoskeletal symptoms such as arthralgia, myalgia, and morning stiffness. SS directly affects the peripheral joints, causing arthralgia in approximately 90% of the patients. Up to 17% arthritis incidence has been reported in SS patients [14, 15]. Arthritis is often symmetrical, intermittent, non-erosive and does not leave deformity. SS mainly affects the metacarpophalangeal joints located in the upper extremity, particularly the metacarpophalangeal joints with non-erosive synovitis, but also affects the wrists, knees, shoulders, and metatarsophalangeal joints [16]. Clinical symptoms of SS are similar to those of rheumatoid arthritis (RA), with the exception of bone erosion, which is very rare in SS [17]. Joint symptoms associated with primary Sjögren's syndrome (pSS) were reported as synovitis, which can mimic rheumatoid arthritis but were distinguished based on the absence of structural damage [18]. Joint involvement may precede the onset of SS in 10–20% of patients, but in a large amount of SS patients (40–50% of the cases) its onset is concurrent with the onset of sicca symptoms [19]. Ultrasonographic US imaging has proven to be of great value in identification of inflammatory synovitis and detection of erosions. Patients with SS were evaluated by ultrasonography(US), the prevalences of synovitis and erosion were found as 21.7% and 34.8%, respectively [20]. The incidence of synovitis in the metacarpophalangeal joints was found as 41.6% [21]. Additionally, a significant correlation was found between the ESSDAI scores of the SS patients whose disease activities were determined according to European League Against Rheumatism (EULAR) Sjogren's Syndrome Disease Activity Index (ESSDAI) and the frequency of synovitis and tenosynovitis detected by US. Based on these findings, it has been suggested that US is a useful method in the evaluation of joint involvement in SS [22]. SS is a complex and heterogeneous disease and has pleomorphic symptoms that can manifest in many different ways [23]. Joint involvement was present in 31.4% of the diagnosed patients, and was manifested as the first symptom in 17% of the patients [24]. Joint involvement is also associated with the presence of many serological markers such as cryoglobulins characterized by extraglandular manifestations, hypergammaglobulinemia, rheumatoid factor (RF) and anti-Sjögren's-syndrome-related antigen A (anti-SSA/Ro) or anti-Sjögren's-syndrome-related antigen B (anti-SSB/La) antibodies [25]. Apart from RA, the highest percetange of RF positivity is seen in SS patients. Approximately 40% of the SS patients, and even a higher percentage if only the SS patients with joint involvement were considered, were found to have RF positivity [25]. Another serological marker associated with arthritic manifestations is anti-citrullinated protein antibodies (ACPAs). ACPAs were found in 5–10% of the SS patients. SS patients with ACPAs were found to have a higher incidence of arthritis than those without ACPAs (43.7% vs. 12.2%). In addition, during a follow-up period of 5–10 years, 43.8% of the SS patients with ACPAs were found to have developed RA [26]. SS patients with ACPAs had a higher tendency to have arthritis and were at a higher risk of developing RA [27]. Joint involvement such as arthralgia and arthritis negatively affects the quality of life in patients with SS [28], creating the need for pharmacological treatment or surgical intervention [29]. It has been emphasized that SS causes higher disease activity scores, which are expressed using scoring systems such as ESSDAI, due to joint involvement, and that joint involvements an important clinical feature in predicting long-term disease outcome in SS [30]. Arthritic manifestations of the SS are mild, but course of SS in association with other diseases varies greatly. However, it is still unclear whether this variation is due to any change directly associated with SS or the combined effect of the disorders accompanying SS. Findings that prove the association of SS with other autoimmune diseases have been demonstrated by epidemiological and genetic studies. SS was accompanied by RA and SLE in 19.5% and 13.6% of the patients, respectively [31]. There are also studies that revealed an epigenetic relationship linking such diseases. In one of these studies, a gene-expression meta-analysis study in respect of RA, SLE and SS, a common gene-expression was identified for these diseases [32]. It is a common concern that such disease combinations will worsen joint problems and adversely affect the course of the disease. The frequency of RA and SS association and the effects of this association on the course of the disease and comorbid conditions, it was found that 31.2% of the RA patients were also diagnosed with SS, and that the coexistence of these two diseases led to higher disease burden, higher disease activity, higher number og comorbidities (hypertension, cardiovascular diseases, malignancy and infections) and a higher degree of erosive changes [33]. Therefore, type of SS, whether it is primary or secondary SS, should be carefully considered and it should be kept in mind that other autoimmune diseases, including RA, can accompany SS during the course of the disease. Thus, the attending clinician should be able to also characterize these other patient populations.

4. Muscles

Muscle pain has been reported in approximately 45–50% of the patients with SS [34]. A thorough literature review reveals that a mild inflammatory myopathy with subclinical or insidious onset has been observed in SS. Generally, this condition manifests itself as muscle pain and proximal muscle weakness. Histopathological examination was performed in SS patients with muscle pain, inflammation was found in 72% of the patients, and signs of degeneration/regeneration (i.e., histological findings of myositis) were found in 47% of the patients along with inflammation [35]. In that regard, it has been shown that although rare, inflammatory muscle diseases (particularly, inclusion body myositis and polymyositis) may be associated with SS. Accordingly, patients with more insidious onset muscle weakness and low muscle enzyme elevations in particular should be suspected of inflammatory muscle diseases. There are case reports, which argued that inflammatory muscle diseases and SS can progress together in patients with SS, and that the possible cause of the co-existence of these diseases is a common autoimmune pathway [36, 37]. Additionally, it was reported that cytosolic 5′-nucleotidase 1A, which is a specific marker for inclusion body myositis, can be detected in approximately 30% of the patients with SS [38]. Given this finding, SS patients with muscle weakness should also be screened for inflammatory muscle diseases.

5. Fibromyalgia and fatigue

Fibromyalgia (FM) is a common disease characterized by widespread chronic body pain, sleep disturbance, weakness, and mood disorders. One of the hypotheses put forward in respect of the formation of FM disease is chronic inflammation. It was reported in many studies that pro-inflammatory cytokines and mediators are higher in patients with FM than in general population [39]. Frequency of FM is high in rheumatic diseases such as ankylosing spondylitis and rheumatoid arthritis. One-third of FM patients (about 33%) with sicca syndrome and/or xerostomia tested positive for Sjögren's syndrome biomarkers [40]. In parallel, association of FM with these autoimmune diseases is common. To give an example, in a recent study, in which the frequency of FM was investigated in patients with inflammatory arthritis (IA), it was found that frequency of FM in IA patients was found to be 15–20% more than the frequency of FM in general population [41]. When it comes to FM and SS, it is seen that these two diseases share common symptoms such as muscle aches, fatigue, and dry mouth and eyes, and there are many studies available in the literature that investigated the relationship between the two. In one of these studies, in which SS-related auto-antibodies in FM patients presented with dry mouth and/or dry eye complaints were investigated, SS biomarkers were found to be positive in 32% of the FM patients. It was suggested by the authors of the study that SS may play a role in the pathophysiology of FM [40]. The comorbidities and clinical symptoms of the two diseases largely overlap. To give a few examples; in a cross-sectional study conducted by Choi et al., which investigated the frequency and clinical effect of FM onpatients with SS, as well as FM frequency, disease activity scales such as Eular Sjogren's Syndrome Patient Reported Index (ESSPRI) and Eular Sjogren Syndrome Disease Activity Index (ESSDAI), and depression in patients with pSS, FM was detected in 31% of the SS patients, and both ESSPRI and ESSDAI scores and depression scores were found to be higher in patients who have both SS with FM, as compared to patients without FM [42]; and in a very recent population-based retrospective-cohort study conducted with 149.706 participants, in which the future risk of developing SS in patients with FM was investigated, patients with FM were found to have a higher risk of developing SS as compared to the control subjects without FM [43]. Given the results of these studies, which suggest that the relationship between SS and FM affects the disease activity and diagnosis, clinicians should consider the two-way relationship between SS and FM in the management of SS or FM. Fatigue is one of the most common symptoms of both SS and FM. The pathophysiology in SS is not fully understood. Patients with SS suffer from sleep disorders as they feel the urge to drink excess water due to dry mouth resulting in sleep disruptions. Fatigue in patients with SS has been associated with the coexistence of sleep disorders and FM. The pathophysiology in SS is not fully understood. Patients with SS suffer from sleep disorders, since they feel the urge to drink excess water due to dry mouth. Fatigue in patients with SS has been associated in the literature with the coexistence of sleep disorders and FM. To give an example; ina large cohort study conducted with 437 patients with pSS, it was found that patients with both FM and pSS manifested significantly more structural, fatigue, and arthralgia symptoms than patients with only pSS [44]. The presence of such symptoms substantially impairs the quality of life in patients with SS. As a matter of fact, it was concluded as a result of a cross-sectional study conducted using questionnaires to measure quality of life of patients with SS that the main determinants of the poor quality of life in patients with SS were pain and fatigue, and that disease activity scores were higher in patients with a high incidence of the said symptoms [45]. For this reason, sleep quality of SS patients should be improved against the symptom of fatigue, which is very common in patients with SS. In addition, in the event that SS accompanied by FM, disease management should be adjusted accordingly.

6. Bones

SS impairs the bone metabolism not just because it is a systemic autoimmune disease, but also because of some other factors associated with SS such as interstitial nephritis, renal tubular acidosis, steroid use, coexistence with other autoimmune diseases, and low vitamin D levels. As is the case with other autoimmune diseases, SS also causes osteoporosis (OP) and osteomalacia (OM). SS plays a role in metabolic bone diseasesby causing variations in the signaling pathways of the Wingless-type (Wnt) and Nf-κB receptor activating factor (RANK), its ligand (RANKL) and osteoprotegerin (OPG) [46]. It has been reported in the pathophysiological studies conducted on the aforementioned signaling pathways that autoimmune diseases such as SS inhibits bone formation, since it reduces the levels of DKK1 (Dickkopf-related protein 1), a protein which is involved in the Wnt pathway and plays a role in bone formation. In addition, it has been shown that the RANKL/RANK/OPG signaling pathway, which features the main osteogenic factors and plays an important role in bone homeostasis, is activated in many autoimmune diseases, and it has been suggested that this leads to bone destruction [47, 48]. Furthermore, the Wnt/b-catenin signaling pathway plays a key role also in the development and regulation of the immune system, in addition to organogenesis and morphogenesis of the exocrine gland. In that regard, it was found in a study by Fernandez-Torres et al. on the genetic polymorphisms of the Wnt/b-catenin signaling pathway in SS patients that some genes associated with the Wnt/b-catenin signaling pathway, such as LRP5 (low-density lipoprotein receptor-related protein 5), FRZB (frizzled related protein), and ADIPOQ (adiponectin), significantly increase the risk of developing SS [49]. In another study, in which impaired bone metabolism in SS was investigated, osteoporosis/osteopenia was detected in ⅔ (two-thirds) of the SS patients. In the same study, it has been shown that DKK1, one of the Wnt signal mediators, was low in SS patients and that this low level is associated with a decrease in bone mineral density, suggesting that Wnt signaling mediators are potentially involved in the pathogenesis of SS [50]. In a case-control study conducted by Pasoto et al. with 71 SS patients and 71 healthy control subjects of matching age, sex, and race, study participants were screened for bone mineral density (BMD), vertebral fracture (VF), and bone microarchitecture (by means of high-resolution peripheral quantitative computer tomography (HR-pQCT)). Consequentially, as compared to the healthy control subjects, it was found that patients with SS had lower mean BMD values in both hip and lumbar vertebrae, and less cortical bone thickness, and that a higher frequency of SS patients (approximately 20% of the SS patients) had VF and significantly impaired bone microarchitecture [51]. Additionally, in the same study, HR-pQCT revealed significant cortical deterioration in SS patients. However, the final assessment on the primary causative factor for osteoporosis (OP) could not be made, due to the fact that all SS patients were on corticosteroid therapy at the time of the study. There is a significant relationship between corticosteroid use and the development of OP. In a recent large-scale cross-sectional study, in which the frequency of OP, risk factors and fragility fractures were investigated in relation to SS, a significant correlation was found between the development of OP and factors such as age, duration of disease, corticosteroid use, presence of anti-La antibodies and ESSDAI scores in patients with SS. Up to 8.5% of the patients with SS were found to have fragility fractures, and a significant correlation was observed between the SS disease duration and age, ESSDAI scores and fragility fracture [52]. Osteomalacia (OM) is a disease characterized by impaired bone mineralization. The development of OM in SS patients has been associated with tubulointerstitial nephritis (TIN) or distal renal tubular acidosis (dRTA). There areseveral studies available in the literature to that effect thathighlight the effect of TIN and dRTA in patients with SS and its contribution to the development of osteomalacia [53, 54]. Patients with OM are susceptible to pseudo-fractures.

Therefore, diagnosis of OM in patients with SS is very important. It is not yet known whether these findings can be directly attributed to the relationship between SS, OP and OM, pathophysiologically. Nevertheless, the direct clinical relevance between these conditions should not be overlooked. The exact mechanism of the effect of SS on bone metabolism has not been determined, but it is obvious that there is a relationship, albeit an indirect one. The part which is not yet clear is whether it is the SS itself or the drugs used for the treatment of SS or the target organs affected in relation thereto are the main factors creating the said effect.

7. Available therapeutic options with possible benefit in musculoskeletal disease

There is no curative treatment for SS, thus the aim in the treatment of SS is to alleviate the symptoms of exocrinopathy and also to get the extraglandular manifestations of the disease under control. Management of SS patients requires a multidisciplinary approach involving collaboration with specialist doctors from different specializations, such as clinical immunologists, rheumatologists, ophthalmologists, otolaryngologists and/or dentists. Centers such as The Sjögren's Foundation, The British Society for Rheumatology, and EULAR publish guidelines for the management of SS [55–57]. Nevertheless, no specific therapeutic goal other than symptomatic relief has been put forward in these guidelines. Fatigue, one of the major symptoms targeted to be relieved by the treatment modalities used for the treatment of SS, is a symptom that significantly impairs quality of life in patients with SS. However, the effectiveness of the current medical treatments used to contain this symptom is still not up to the level. Available guidelines suggest regular physical activity as the best approach to improve fatigue [56, 57]. Hydroxychloroquine (HCQ) is usually recommended as the first-line treatment for musculoskeletal pain relief. On the other hand, methotrexate (MTX) is recommended to be used as a stand-alone medication or in combination with HCQ in patients who do not respond to HCQ, and particularly in those with severe inflammatory arthritis [56]. In cases where the combination of HCQ and MTX has proven ineffective in the treatment of inflammatory musculoskeletal symptoms, alternative options such as use of corticosteroids, leflunomide, sulfasalazine, azathioprine, cyclosporine, or biologic drugs may be considered [55, 57].

Author details

Ridvan İşik* and Ferhat Ege
Hatay Training and Research Hospital, Hatay, Turkey

*Address all correspondence to: dr.ridvanisik@gmail.com

References

[1] Henrik S. Zur Kenntnis der keratoconjunctivitis sicca. Keratitis filiformis bei Hypofunktion der Tränendrüsen [On knowledge of keratoconjunctivitis sicca. Keratitis filiformis due to lacrimal gland hypofunction]. Acta Ophthalmologica. 1933;**2**:1-151

[2] Brito-Zeron P, Theander E, Baldini C, et al. Early diagnosis of primary Sjogren's syndrome: EULAR-SS task force clinical recommendations. Expert Review of Clinical Immunology. 2016;**12**(2): 137-156

[3] Qin B, Wang J, Yang Z, Yang M, Ma N, Huang F, et al. Epidemiology of primary Sjögren's syndrome: A systematic review and meta-analysis. Annals of the Rheumatic Diseases. 2015;**74**(11):1983

[4] Brito-Zeron P, Acar-Denizli N, Zeher M, et al. Influence of geolocation and ethnicity on the phenotypic expression of primary Sjogren's syndrome at diagnosis in 8310 patients: A cross-sectional study from the big data Sjogren project consortium. Annals of the Rheumatic Diseases. 2017;**76**(6): 1042-1050

[5] Saraux A, Pers JO, Devauchelle-Pensec V. Treatment of primary Sjogren syndrome. Nature Reviews Rheumatology. 2016;**12**:456-471

[6] Shimizu T, Nakamura H, Kawakami A. Role of the innate immunity signaling pathway in the pathogenesis of Sjögren's syndrome. International Journal of Molecular Sciences. 2021;**22**(6):3090

[7] Brito-Zeron P, Baldini C, Bootsma H, et al. Sjogren syndrome. Nature Reviews. Disease Primers. 2016;**2**:16047

[8] Bombardieri M, Argyropoulou OD, Ferro F, et al. One year in review 2020: Pathogenesis of primary Sjogren's syndrome. Clinical and Experimental Rheumatology. 2020;**126**(4):3-9

[9] Chivasso C, Sarrand J, Perret J, Delporte C, Soyfoo MS. The involvement of innate and adaptive immunity in the initiation and perpetuation of Sjogren's syndrome. International Journal of Molecular Sciences. 2021;**22**(2):658

[10] Vıtalı C, Dolcıno M, Del Papa N, et al. Gene expression profiles in primary Sjögren's syndrome with and without systemic manifestations. ACR Open Rheumatology. 2019;**1**:603-613

[11] Imgenberg-kreuz J, AlmlÖF JC, Leonard D, et al. Shared and unique patterns of DNA methylation in systemic lupus erythematosus and primary Sögren's syndrome. Frontiers in Immunology. 2019;**10**:1686

[12] Davıes R, Sarkar I, Hammenfors D, et al. Single cell based phosphorylation profiling identifies alterations in toll-like receptor 7 and 9 signaling in patients with primary Sjögren's syndrome. Frontiers in Immunology. 2019;**10**:281

[13] Luo S, Zhu R, Yu T, et al. Chronic inflammation: A common promoter in tertiary lymphoid organ neogenesis. Frontiers in Immunology. 2019;**10**:2938

[14] Baldini C, Pepe P, Quartuccio L, et al. Primary Sjogren's syndrome as a multi-organ disease: Impact of the serological profile on the clinical presentation of the disease in a large cohort of Italian patients. Rheumatology (Oxford, England). 2014;**53**(5):839-844

[15] Ramos-Casals M, Brito-Zerón P, Solans R, et al. Systemic involvement in

primary Sjogren's syndrome evaluated by the EULAR-SS disease activity index: Analysis of 921 Spanish patients (GEAS-SS Registry). Rheumatology (Oxford, England). 2014;**53**(2):321-331

[16] Amezcua-Guerra LM, Hofmann F, Vargas A, et al. Joint involvement in primary Sjögren's syndrome: An ultrasound "target area approach to arthritis". BioMed Research International. 2013;**2013**:640265

[17] Jacques T, Sudoł-Szopińska I, Larkman N, O'Connor P, Cotten A. Musculoskeletal manifestations of non-RA connective tissue diseases: Scleroderma, systemic lupus erythematosus, Still's disease, dermatomyositis/polymyositis, sjögren's syndrome, and mixed connective tissue disease. Seminars in Musculoskeletal Radiology. 2018;**22**(2):166-179

[18] Mirouse A, Seror R, Vicaut E, et al. Arthritis in primary Sjögren's syndrome: Characteristics, outcome and treatment from French multicenter retrospective study. Autoimmunity Reviews. 2019; **18**(1):9-14

[19] Vitali C, Del Papa N. Pain in primary Sjögren's syndrome. Best Practice & Research. Clinical Rheumatology. 2015;**29**(1):63-70

[20] Lei L, Morgan S, Ntatsaki E, Ciurtin C. Comparative assessment of hand joint ultrasound findings in symptomatic patients with systemic lupus erythematosus and Sjögren's syndrome: A pilot study. Ultrasound in Medicine & Biology. 2019;**45**(2):452-460

[21] Riente L, Scirè CA, Delle Sedie A, et al. Ultrasound imaging for the rheumatologist. XXIII. Sonographic evaluation of hand joint involvement in primary Sjögren's syndrome. Clinical and Experimental Rheumatology. 2009;**27**(5):747-750

[22] Guedes LKN, Leon EP, Bocate TS, Bonfigliolli KR, Lourenço SV, Bonfa E, et al. Characterizing hand and wrist ultrasound pattern in primary Sjögren's syndrome: A case-control study. Clinical Rheumatology. 2020;**39**(6):1907-1918

[23] Leone MC, Alunno A, Cafaro G, et al. The clinical spectrum of primary Sjögren's syndrome: Beyond exocrine glands. Reumatismo. 2017;**69**(3):93-100

[24] Fauchais AL, Ouattara B, Gondran G, et al. Articular manifestations in primary Sjögren's syndrome: Clinical significance and prognosis of 188 patients. Rheumatology (Oxford, England). 2010;**49**(6): 1164-1172

[25] García-Carrasco M, Ramos-Casals M, Rosas J, et al. Primary Sjögren syndrome: Clinical and immunologic disease patterns in a cohort of 400 patients. Medicine (Baltimore). 2002;**81**(4): 270-280

[26] Payet J, Belkhir R, Gottenberg JE, et al. ACPA-positive primary Sjögren's syndrome: True primary or rheumatoid arthritis-associated Sjögren's syndrome? RMD Open. 2015;**1**(1):26-44

[27] Molano-González N, Olivares-Martínez E, Anaya JM, Hernández-Molina G. Anti-citrullinated protein antibodies and arthritis in Sjögren's syndrome: A systematic review and meta-analysis. Scandinavian Journal of Rheumatology. 2019;**48**(2):157-163

[28] Cornec D, Devauchelle-Pensec V, Mariette X, et al. Severe health-related quality of life impairment in active primary Sjögren's syndrome and patient-reported outcomes: Data from a large therapeutic trial. Arthritis Care & Research. 2017;**69**(4):528-535

[29] Chu LL, Cui K, Pope JE. Meta-analysis of treatment for primary

Sjögren's syndrome. Arthritis Care & Research. 2020;**72**(7):1011-1021

[30] Quartuccio L, Gandolfo S, Zabotti A, Zandonella Callegher S, Fabro C, De Vita S. Articular and peripheral nervous system involvement are linked to the long-term outcome in primary Sjögren's syndrome: The relevance of single organ manifestations rather than a composite score as predictors. Frontiers in Immunology. 2019;**10**:1527

[31] Alani H, Henty JR, Thompson NL, Jury E, Ciurtin C. Systematic review and meta-analysis of the epidemiology of polyautoimmunity in Sjögren's syndrome (secondary Sjögren's syndrome) focusing on autoimmune rheumatic diseases. Scandinavian Journal of Rheumatology. 2018;**47**(2):141-154

[32] Toro-Domínguez D, Carmona-Sáez P, Alarcón-Riquelme ME. Shared signatures between rheumatoid arthritis, systemic lupus erythematosus and Sjögren's syndrome uncovered through gene expression meta-analysis. Arthritis Research & Therapy. 2014;**16**(6):489

[33] Harrold LR, Shan Y, Rebello S, et al. Prevalence of Sjögren's syndrome associated with rheumatoid arthritis in the USA: An observational study from the Corrona registry. Clinical Rheumatology. 2020;**39**(6):1899-1905

[34] Tishler M, Barak Y, Paran D, Yaron M. Sleep disturbances, fibromyalgia and primary Sjögren's syndrome. Clinical and Experimental Rheumatology. 1997;**15**:71-74

[35] Lindvall B, Bengtsson A, Ernerudh J, Eriksson P. Subclinical myositis is common in primary Sjögren's syndrome and is not related to muscle pain. The Journal of Rheumatology. 2002;**29**(4):717-725

[36] Kanellopoulos P, Baltoyiannis C, Tzioufas AG. Primary Sjögren's syndrome associated with inclusion body myositis. Rheumatology (Oxford, England). 2002;**41**(4):440-444

[37] Migkos MP, Sarmas I, Somarakis GA, Voulgari PV, Tsamis KI, Drosos AA. Sjögren syndrome associated with inflammatory muscle diseases. Mediterranean Journal of Rheumatology. 2018;**29**(2):92-96

[38] Rietveld A, van den Hoogen LL, Bizzaro N, et al. Autoantibodies to cytosolic 5'-nucleotidase 1A in primary Sjögren's syndrome and systemic lupus erythematosus. Frontiers in Immunology. 2018;**9**:1200

[39] Mendieta D, De la Cruz-Aguilera DL, Barrera-Villalpando MI, et al. IL-8 and IL-6 primarily mediate the inflammatory response in fibromyalgia patients. Journal of Neuroimmunology. 2016;**290**:22-25

[40] Applbaum E, Lichtbroun A. Novel Sjögren's autoantibodies found in fibromyalgia patients with sicca and/or xerostomia. Autoimmunity Reviews. 2019;**18**(2):199-202

[41] Zhao SS, Duffield SJ, Goodson NJ. The prevalence and impact of comorbid fibromyalgia in inflammatory arthritis. Best Practice & Research. Clinical Rheumatology. 2019;**33**(3):101423

[42] Choi BY, Oh HJ, Lee YJ, Song YW. Prevalence and clinical impact of fibromyalgia in patients with primary Sjögren's syndrome. Clinical and Experimental Rheumatology. 2016;**34** (2 Suppl 96):S9-S13. Epub 2015 Aug 27

[43] Gau SY, Leong PY, Lin CL, Tsou HK, Wei JC. Higher risk for Sjögren's syndrome in patients with fibromyalgia: A nationwide population-based Cohort study. Frontiers in Immunology. 2021;**12**:640618

[44] Torrente-Segarra V, Corominas H, Sánchez-Piedra C, et al. Fibromyalgia prevalence and associated factors in primary Sjögren's syndrome patients in a large cohort from the Spanish Society of Rheumatology registry (SJOGRENSER). Clinical and Experimental Rheumatology. 2017;**35**

[45] Dias LH, Miyamoto ST, Giovelli RA, de Magalhães CIM, Valim V. Pain and fatigue are predictors of quality of life in primary Sjögren's syndrome. Advances in Rheumatology. 2021;**61**(1):28

[46] Skarlis C, Palli E, Nezos A, Koutsilieris M, Mavragani CP. Study of the incidence of osteoporosis in patients with Sjögren's syndrome (pSS) and investigation of activation of the RANKL/RANK and osteoprotegerin (OPG) system. Mediterranean Journal of Rheumatology. 2018;**29**(4):224-227

[47] Baron R, Kneissel M. WNT signaling in bone homeostasis and disease: From human mutations to treatments. Nature Medicine. 2013;**19**(2):179-192

[48] Leibbrandt A, Penninger JM. RANK/RANKL: Regulators of immune responses and bone physiology. Annals of the New York Academy of Sciences. 2008;**1143**:123-150

[49] Fernández-Torres J, Pérez-Hernández N, Hernández-Molina G, et al. Risk of Wnt/β-catenin signalling pathway gene polymorphisms in primary Sjögren's syndrome. Rheumatology (Oxford, England). 2020;**59**(2): 418-425

[50] Gravani F, Papadaki I, Antypa E, et al. Subclinical atherosclerosis and impaired bone health in patients with primary Sjogren's syndrome: Prevalence, clinical and laboratory associations. Arthritis Research & Therapy. 2015; **17**(1):99

[51] Pasoto SG, Augusto KL, Alvarenga JC, Takayama L, Oliveira RM, Bonfa E, et al. Cortical bone density and thickness alterations by high-resolution peripheral quantitative computed tomography: Association with vertebral fractures in primary Sjögren's syndrome. Rheumatology (Oxford, England). 2016;**55**(12):2200-2211

[52] Salman-Monte TC, Sanchez-Piedra C, Fernandez Castro M, et al. Prevalence and factors associated with osteoporosis and fragility fractures in patients with primary Sjögren syndrome. Rheumatology International. 2020;**40**(8): 1259-1265

[53] Geng Y, Zhao Y, Zhang Z. Tubulointerstitial nephritis-induced hypophosphatemic osteomalacia in Sjögren's syndrome: A case report and review of the literature. Clinical Rheumatology. 2018;**37**(1):257-263

[54] Nagae H, Noguchi Y, Ogata S, et al. Osteomalacia secondary to renal tubular acidosis due to Sjögren's syndrome: A case report and review of the literature. CEN Case Reports. 2012;**1**(2):123-127

[55] Vivino FB, Carsons SE, Foulks G, et al. New treatment guidelines for Sjogren's disease. Rheumatic Diseases Clinics of North America. 2016;**42**(3): 531-551

[56] Price EJ, Rauz S, Tappuni AR, et al. The British Society for Rheumatology guideline for the management of adults with primary Sjögren's syndrome. Rheumatology. 2017;**56**(10):1643-1647

[57] Ramos-Casals M, Brito-Zeron P, Bombardieri S, et al. EULAR recommendations for the management of Sjogren's syndrome with topical and systemic therapies. Annals of the Rheumatic Diseases. 2020;**79**(1):3-18

Chapter 3

Musculoskeletal Abnormalities Caused by Cystic Fibrosis

Mark Lambrechts

Abstract

Cystic Fibrosis (CF) can affect all organs of the human body including the musculoskeletal system. Although the musculoskeletal aspects of CF are less commonly studied, fractures (predominantly spinal), muscle injuries, and joint pain are more commonly seen in the CF population compared to the general public due to their lower bone mineral density, dysfunctional skeletal muscle, and elevated levels of pro-inflammatory cytokines. Additionally, due to elevated levels of inflammation in the CF population diagnosis of musculoskeletal injuries can be difficult to pinpoint. As treatment for CF evolves, an increased understanding of how CF affects the musculoskeletal system is imperative. We will discuss the orthopedic aspects of CF and provide potential insights into the future direction of orthopedic care in the CF population.

Keywords: cystic fibrosis, musculoskeletal, spine, arthropathy, fracture, bisphosphonates, allele specific drugs, cytokines, bone mineral density

1. Introduction

Cystic Fibrosis (CF) is an autosomal recessive disorder causing loss of function of the cystic fibrosis transmembrane conductance regulator (CFTR) gene. The most common gene alteration present in CF patients is a deletion of phenylalanine at position 508 (ΔF508). Fortunately, researchers and pharmaceutical companies have produced allele-specific drugs targeting CF genetic mutations. These work through multiple potential mechanisms, but allele potentiation (ivacaftor) and allele correction (tezacaftor) are some of the more promising therapeutics to date [1, 2].

Historically, patients with any combination of CFTR gene mutations could invariably expect a progressive course of worsening respiratory and endocrine function, resulting in irreversible and severe lung and pancreas damage. However, the introduction of allele-specific drugs has had a profound impact on improving both the length and quality of life in CF patients [3]. Given the improved life expectancy, orthopedic providers can expect to see a resultant increase in the proportion of patients with CF.

Throughout this chapter, we will discuss the three most likely scenarios for orthopedic consultation in the CF population: bone health/fracture, muscle and soft tissue dysfunction, and joint pain and arthralgia. In order to understand each of these topics appropriately, each subtopic will be prefaced by an in-depth introduction to the basic science causing the musculoskeletal pathology with subsequent detailed management of the disease.

1.1 Cystic fibrosis- related bone disease (CFBD): understanding the abnormal molecular pathway

Healthy bone undergoes continuous remodeling – bone resorption is mediated by osteoclasts through the RANK-RANKL (receptor activator of nuclear factor kappa-β ligand) pathway. Meanwhile, bone deposition occurs through activation of osteoblasts, which signal through the WNT-β-catenin pathway [4]. Osteoblasts also secrete osteoprotegrin, which binds to RANKL thus limiting activation of osteoclasts. In this manner, osteoblasts and osteoclasts are in delicate balance, and their goal is to optimize bone mineral density (BMD) while minimizing unnecessary storage of essential nutrients via reorganization of the boney trabecular microarchitecture. Bones under continual heavy loads (stress) or tension (strain) will adapt to the increase in forces imparted to the bone, while bones undergoing less frequent loading, will have a resultant leach of essential nutrients, thus allowing each bone to maximize its function (Wolff's Law) [5].

Mounting research is focused on improving our understanding of osteoblast and osteoclast function in CF patients. Emerging evidence indicates CF patients with pulmonary exacerbations caused by underlying indolent lung infections, have elevated cytokine levels [6]. The systemic increase in pro-inflammatory cytokines during these CF "flare-ups" leads to formation and activation of osteoclasts, resulting in bone resorption [7]. Additionally, the ΔF508 phenotype is known to promote RANKL production, which is normalized with the allele specific drug ivacaftor [8–10] and the drug miglustat [11].

Cystic Fibrosis not only increases osteoclast activity, but it also detrimentally affects osteoblast function and uncouples osteoblast–osteoclast homeostatsis leading to severe trabecular and cortical bone osteopenia through net bone mineral resorption [12]. Diminished ΔF508 osteoblast activity is thought to contribute to poor COX-2, PGE_2, and osteoprotegrin expression [13]. One potential therapeutic to mitigate poor bone quality and inhibited fracture healing is ivacaftor, which works through allele potentiation of the ΔF508-CFTR channel and channel optimization returns osteoblast function to 85% of normal [14]. This effectively increases systemic levels of cyclooxygenase-2 (COX-2) and PGE_2, which are integral for effective bone maturation and fracture healing [13].

Translational research performed in mouse models indicates that $Cftr^{-/-}$ mice have 50% less cortical bone width, thinner and less plentiful trabeculae and greater trabecular separation compared to normal mice [15]. Trabecular bone formation was also decreased by 92% in $Cftr^{-/-}$ mice likely due to the density of osteoclasts near the cortical surface [15]. Skeletal growth also appears to be hindered by 40% in $Cftr^{-/-}$ mice, due to a reduction in the hypertrophic zone of the growth plate [15]. The combination of these findings results in smaller bones that have decreased thickness and strength compared to normal bones [16]. This results in CF bones being able to tolerate less load to failure, which increases fracture risk [17]. It is believed that issues with poor BMD and increased fracture risk manifest prior to age six with no further significant worsening of BMD until after adolescence [18]. However, exercise programs in pre-adolescent children may increase BMD by 7% and these programs should be routinely implemented in all children with CF [19].

Aside from the poor bone quality imparted from phenotypic alterations by the CFTR gene, additional causes of lower BMD in CF patients include female sex, cystic fibrosis related diabetes (CFRD), vitamin D malabsorption, malnutrition, pancreatic insufficiency, exogenous corticosteroid use, chronic inflammation/chronic

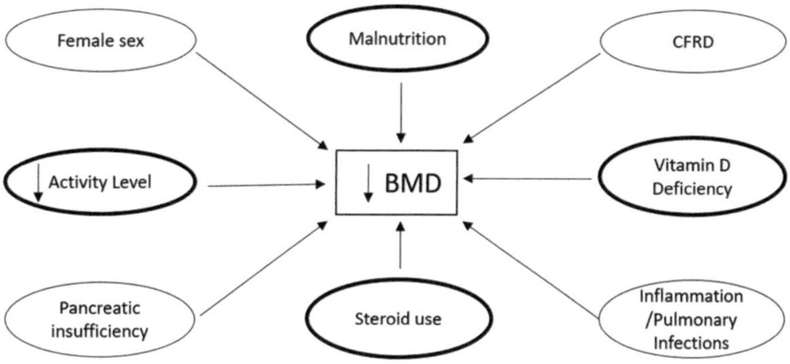

Figure 1.
*Risk factors for decreased bone mineral density (BMD) in patients with cystic fibrosis. * bold ovals indicate high quality evidence for risk factor contribution to decreased BMD in cystic fibrosis patients, while non-bolded ovals indicate poor quality evidence of contribution to diminished BMD.*

pulmonary infections, and decreased activity levels (**Figure 1**). It should be noted that there is a paucity of well-designed studies demonstrating female sex [15, 20, 21], pancreatic [21, 22], CFRD [22, 23], and chronic inflammation/pulmonary infections [18, 24] have a significant effect on BMD. However, sufficient evidence does appear to suggest decreased activity levels, malnutrition, exogenous corticosteroid use, and vitamin D deficiency or malabsorption are potentially controllable risk factors that lead to a significant reduction in BMD [18, 21, 22, 25].

The net bone resorption in CF patients can be evaluated through serum or urinary analysis of type I collagen N-telopeptides, free deoxypyridinoline, and alkaline phosphatase. Increased levels of each of these molecular markers is common in CF patients, which may be further elevated by increased serum parathyroid hormone (PTH) levels and diminished serum 25-hydoxyvitamin D levels in these patients [26]. PTH and 25-hydoxyvitamin D create a positive feedback loop signaling the body to continue to resorb bone and increase serum calcium levels [27]. The combination of elevated osteoclast function and diminished osteoblast function results in early osteopenia or osteoporosis, which are diagnosed in 23.5% and 38% of the adult CF population, respectively [28].

1.2 Is there a role for bisphosphonates to improve bone health for CF patients?

Vitamin D deficiency affects up to 90% of patients with CF [29]. This is predominantly caused by poor systemic vitamin D absorption since CF patients absorb less than 50% of the dietary vitamin D absorbed by normal patients and 20% of CF patients have no measurable vitamin D absorption [30]. For this reason, the Cystic Fibrosis Foundation recommends each CF patient be given a daily prescription of vitamin D_3 to maintain baseline 25-hyroxyvitamin D levels at a minimum of 30 ng/ml [31]. However, even with sufficient vitamin D levels, patients with CF should undergo a dual energy X-ray absorptiometry (DEXA) scan at the age of 18 [26]. Repeat DEXA scans should occur every five years if the BMD is normal but repeat testing should occur every 2–4 years if the DEXA scan is between the ranges of <−1 and > −2 (osteopenia range) [26].

Current recommendations for initiation of bisphosphonate administration can be seen in **Table 1** [32]. In patients who are started on an oral or intravenous (IV) course of bisphosphonates, an approximate 3–6% increase in BMD can be expected at

Adults	Children/Adolescents
Scheduled for or have previously had an organ transplant AND have a BMD Z or T score of ≤ −1.5	Scheduled for or have previously had an organ transplant AND have a BMD Z or T score of ≤ −2
Utilization of systemic glucocorticoids for longer than 3 months AND have a BMD Z or T score of ≤ −1.5	Utilization of systemic glucocorticoids for longer than 3 months AND have a BMD Z or T score of ≤ −2 OR have a low energy mechanism fracture
Sustain a low-energy mechanism fracture while on a course of systemic glucocorticoids	Sustain or have a history of fracture AND have a BMD Z or T score of ≤ −2
Sustain a low-energy mechanism fracture OR have a femoral neck BMD Z or T score of ≤ −2 AND have more than 4% BMD loss per year	About to start a prolonged course of systemic glucocorticoids AND have a BMD Z or T score of ≤ −2

Table 1.
Indications for initiation of oral or IV bisphosphonates in the adult or pediatric cystic fibrosis population.

1-year in the lumbar spine and femoral neck [33–35]. However, IV bisphosphonates are associated with severe bone pain and flu-like symptoms that should be discussed with patients prior to initiation [33, 36]. Additionally, atypical femur fractures and jaw osteonecrosis are risk factors for long-term treatment with bisphosphonates [37]. Clinicians should also keep in mind that BMD increases after bisphosphonate administration, but there is currently no evidence to support bisphosphonates ability to reduce the likelihood of sustaining a lower extremity or spine fracture [36].

Historically, there has been concern about fracture healing during bisphosphonate use. While there does appear to be a delay in fracture callus reorganization (immature woven bone is not replaced as quickly with mature lamellar bone), bisphosphonates do not inhibit fracture callus formation [38]. This type of fracture healing is referred to as secondary bone healing or endochondral ossification (examples of this type of healing include patients placed in a cast, surgery which involves an intramedullary nail/rod, or a bridge plate where the bone is not compressed together). While this type of fracture healing appears to be less affected by bisphosphonates, animal models suggest primary fracture healing (e.g. compression plating) is affected by bisphosphonates and causes lower BMD at the fracture site, decreases load to failure, and increases the presence of cartilaginous tissue at the fracture site [38]. However, clinical studies have not found any evidence to support a significant difference in fracture healing based on the administration of bisphosphonates, so at this time CF patients should be allowed to continue taking bisphosphonates even in the presence of a fracture [39, 40].

1.3 Predisposition to appendicular skeletal fractures: fracture management

Cystic Fibrosis predisposes patients to fracture even with minimal or no traumatic etiology due to their low BMD. Appendicular skeleton fractures occur at a rate of approximately 20% [28]. Notably, one case report described a femur fracture in an adolescent male baseball player who was running and had no associated trauma. In this instance, the femur fracture healed after treatment with an intramedullary nail; however, the patient then had an atraumatic contralateral femur fracture months later that was treated in the same manner [41]. There have also been reports of a unilateral femoral neck fracture in a 25-year-old male without associated trauma and bilateral femoral neck fractures in a 34-year-old male after a grand mal seizure [42, 43]. The 25-year-old patient had severe osteoporosis and was treated with internal fixation, while the 34-year-old was treated with bilateral bipolar hemiarthroplasties [42, 43]. It should be noted that in non-CF patients, the femoral neck fractures would have been treated with fracture fixation instead of hip replacement. Since CF patients are now frequently treated with allele-specific drugs, there is no evidence to indicate CF patients should have fractures managed any different from patients without a diagnosis of Cystic Fibrosis. In instances of delayed fracture healing, subcutaneous teriparatide may be another effective tool to promote fracture healing, although there is poor quality evidence to support this management [44].

1.4 Spinal fractures and CF-related spine disease

Up to 94% of CF patients have back pain with potential etiologies often multifactorial, but they include muscle weakness, rib fracture, scoliosis, spinous process bursitis, and vertebral fracture [45–47]. A significantly higher rate of vertebral fractures are identified in CF patients with an incidence of approximately 14%, but interestingly BMD and the risk of vertebral fracture is not correlated [28, 48]. Vertebral fractures often result in vertebral wedging, which progresses to structural kyphosis if wedging is greater than 15% [45]. Since vertebral wedging is typically minimal, only 8% of pediatric patients develop structural kyphosis, with the rate nearly doubling in adult patients [45, 49]. This may be due to the increased rates of muscle weakness and osteopenia/osteoporosis as patients age [50].

Additional considerations for spinal pain include spinal process bursitis, which may be caused by improperly fitting high frequency chest wall oscillation devices (vest) [46]. In these instances, the bursitis should be managed expectantly as it will resolve without surgical intervention after appropriate vest adjustment [46]. Another consideration of spinal pain is idiopathic scoliosis, which is more prevalent in the CF population and typically manifests as a short mid thoracic curve [51]. Idiopathic scoliosis in CF patients is often treated non-surgically with bracing [47]. However, in skeletally immature patients with curve progression to 50 degrees the scoliosis should be managed with surgical correction [52].

1.5 Muscular/soft tissue dysfunction: is there a molecular basis for muscular dysfunction?

Diminished muscle mass and force is a common affliction of CF patients and lower extremity muscles are frequently affected to a greater degree than the upper extremity [53]. Theories abound as to the potential causes of muscle weakness and include elevated cytokine (IL-6) levels, low vitamin D levels, corticosteroids, presence of the ΔF508 phenotype, altered CFTR function in the sarcoplasmic reticulum, muscle disuse, and poor pulmonary function (**Figure 2**) [54–57]. The reality is the cumulative effect of each of these mechanisms contributes to decreased muscle mass since each theory is intertwined.

One of the more prominent theories for muscle dysfunction in CF patients includes abnormal function of the CFTR chloride channels in the sarcoplasmic reticulum, which results in inappropriate regulation of calcium homeostasis [57]. Since calcium is essential for muscle depolarization, dysregulation of these channels may lead to muscle mass loss, early fatigue, and generalized weakness. This mechanism was further evaluated in human and mice diaphragms, where intracellular calcium was significantly elevated after muscle depolarization in the presence of an inflammatory environment [54]. In the presence of *Pseudomonas aeruginosa*,

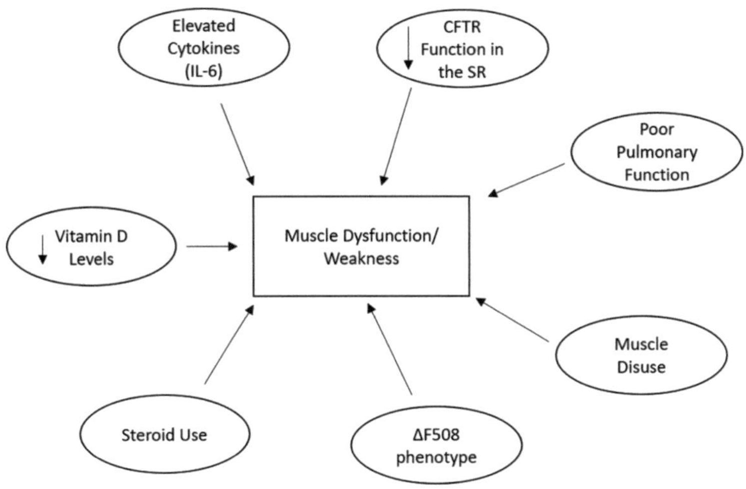

Figure 2.
*Potential contributors to muscle dysfunction and muscle weakness in cystic fibrosis patients. * CFTR = cystic fibrosis transmembrane conductance regulator; SR = sarcoplasmic reticulum.*

elevated pro-inflammatory cytokines were overexpressed and E3 ubiquitin ligases were upregulated (these are often identified in muscle atrophy) resulting in a significant decrease in muscle force generation [54]. Separate research has also identified elevated IL-6 levels and ΔF508 are correlated with diminished muscle mass [56].

The net loss of peripheral muscle strength has continuously been associated with poor pulmonary function [50, 58]. When decreased muscle strength is coupled with poor anaerobic capacity – evidenced by the decreased oxygenation of muscles due to the suboptimal V_O^2 max – [59, 60] early lactic acid accumulation may occur [58]. This can result in muscle disuse through early muscle fatigue. Additional contributing factors include low vitamin D levels, which have been linked to poor peripheral muscle strength although the exact mechanistic role linking vitamin D and muscle weakness is incompletely understood [61]. Luckily, muscle atrophy may not be permanent. Ivacaftor appears to independently increase fat free mass by one kilogram and significantly increases lung function, which may lead to significant long-term improvements in CF patients' health, endurance, and muscle function [62, 63].

Muscle dysfunction may also be linked to low-energy muscle injuries. A single case report identified an adolescent CF patient with an atraumatic mid-substance muscle rupture caused by running during a basketball game. Management of these injuries are typically non-surgical with gradual return to sport [46]. Similar to bony injuries, muscle injuries in the CF population should be treated comparably to muscle injuries in the general population. Future research is necessary to improve our understanding of muscle dysfunction in CF patients and to identify if allele specific drugs are effective at reducing CF patient's predisposition to muscle atrophy and subsequent muscle rupture.

1.6 Weight and aerobic training: improving quality of life

Baseline muscle weakness is present in CF patients, which is linked to decreased quality of life [64]. Therefore, multiple studies have explored the effect of exercise on improvements in strength, endurance, and quality of life [65–67]. An eight-week resistance and aerobic training program demonstrated improved strength, pulmonary function, and % fat free mass [68]. Similar results were found comparing CF patients who participated versus those who did not in a six-month weight training program, which demonstrated effectiveness of the program at increasing muscle size, strength, and overall weight gain [66]. Further, a combined home exercise program including aerobic exercise and resistance training resulted in improved weight gain and quality of life [65]. Respiratory muscle endurance training has also been found to improve both respiratory endurance and exercise endurance [67]. Based on the overwhelming positive effects of exercise on lung and peripheral muscle endurance, consensus guidelines on aerobic activity and resistance training has been established [69]. These guidelines recommend that children and adolescents with CF should engage in at least moderate intensity exercise for 60 minutes per day, while adults should ideally participate in at least moderate intensity exercise for 300 minutes per week [69]. Adults should also participate in upper body, lower body, and trunk resistance training 3–5 times per week with 1–3 sets of 8–15 repetitions. The weight should be based on 70% of maximum weight. In fact, patients with severe CF may maximally benefit from resistance training since they may struggle to have the aerobic capacity for prolonged endurance training [69].

1.7 Joint pain: cytokines effect on the musculoskeletal system

Arthropathy is commonly identified in CF patients at a rate of ranging from 8.5 to 29% [70–72]. Two main forms of CF-related arthropathy exist: cystic fibrosis related arthropathy (CFA) and hypertrophic osteoarthropathy (HOA), although lesser forms of arthropathy have also been described including fluoroquinolone associated arthropathy and an elevated incidence of rheumatoid arthritis [72]. Although no definitive pathway has established the causality of arthropathy in the CF population, risk factors appear to include pulmonary exacerbations with *Pseudomonas* or *Aspergillus*, female gender, older age, serum levels of IgG, CFRD, pancreatic insufficiency, greater number of hospitalizations, and sinusitis (**Figure 3**) [71, 72]. One potential theory linking many of these risk factors with arthropathy is the associated increase in cytokines, especially IL-6, IL-8, and TNF-α [73]. A rapid increase in these cytokines is often associated with pulmonary exacerbations and joint pain is a common additional complaint during these CF "flare-ups" [72]. Further, these pro-inflammatory cytokines are also more prevalent in non-CF patients with radiographic evidence of osteoarthritis and are one potential mechanism that may potentiate joint degeneration [74]. Arthropathy can typically be managed with non-steroidal anti-inflammatory drugs (NSAIDs), which are the first line of treatment, but if corticosteroids are administered due to the underlying pulmonary exacerbation, they may be effective at treating the associated arthropathy [75].

Patients with CF commonly have elevated levels of inflammatory markers including CRP and ESR [76, 77]. The combination of elevated inflammatory markers and a propensity to develop joint pain makes consults for potential septic arthritis more likely for orthopedists. Therefore, physicians need to carefully evaluate the patient's cause of joint pain. Although arthrocentesis may be necessary to rule out septic arthritis for disabling joint pain, inflammatory markers are typically significantly elevated with septic arthritis while they may only be marginally elevated in cases of CFA or HOA [78]. A meticulous physical exam aimed at identifying limited passive range of motion and an inability to ambulate on the affected joint may further distinguish septic arthritis from either CFA or HOA.

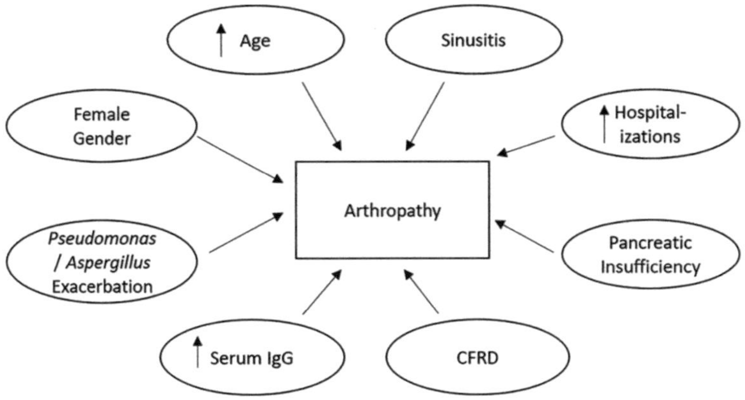

Figure 3.
Potential risk factors contributing to arthropathy episodes in cystic fibrosis patients.

1.8 Cystic fibrosis-related Arthropathy

CFA is the more common form of joint related pain with typical age of onset of approximately 13 years [79]. Although the incidence of CFA ranges from 8.5 to 29%, as the CF population continues to have a longer life span, the expected number of patients with CFA is expected to grow tremendously [71]. As such, practitioners should be cognizant of the symptoms of CFA and should systematically differentiate it from HOA and a septic joint.

CFA has distinct symptoms including, but not limited to, short bursts of recurring episodes of joint pain, fevers, joint swelling, and a rash that resembles erythema nodosum. However, pain and swelling are the most commonly identified forms of CFA and they typically present in the small joints of the hands, although knees and shoulders are also commonly affected [72, 80]. The combination of joint pain, joint swelling, overlying joint reddening, and severe pain causing loss of function is only present in around 13% of patients, allowing CFA to typically be easily differentiated from septic arthritis [72]. Further, CFA is often seen in the setting of oligo- or polyarthritis, which is uncommonly seen in patients with septic arthritis [72]. The onset of CFA symptoms occurs in less than 24 hours but the pain is often limited to four days after initiation of NSAIDs. After symptom resolution, the patient typically has no pain, but the arthropathy typically returns at variable, seemingly random time points. Further, there is often no evidence of radiographic abnormalities if x-rays are taken of the involved joint [79]. Although less commonly associated with pulmonary exacerbations, elevated systemic cytokines may exacerbate CFA [80].

1.9 Hypertrophic osteoarthropathy

HOA is characterized by a combination of medical conditions including periostitis of long bones, digital clubbing, and severe joint arthropathy with or without synovial effusions. Radiographs can help differentiate HOA from CFA due to characteristic periosteal elevation on the distal aspect of the tubular bones [81]. HPOA is also more commonly seen after initiation of a pulmonary exacerbation [82]. CF patients presenting with HOA also typically present with polyarthralgia, which may allow physicians to differentiate it from septic arthritis.

Two main pathways have been proposed for the underlying cause of HOA: humoral and vagal. The humoral pathway has two subtypes: (1) elevated cytokine levels and hypoxemia, which produces hypoxia-inducing factors including VEGF and PDGF or (2) lung arteriovenous [83]. VEGF and PDGF induce proliferation of the endothelial smooth muscle and vascular permeability resulting in angiogenesis. VEGF also stimulates osteoblast and osteoclast induction and the combination of these effects ultimately results in subperiosteal collagen deposition leading to periosteal elevation of the distal portion of the tubular bones [83]. The periosteal reaction seen in this condition results in the variable pain responses seen in these patients. From a mechanistic standpoint, this pathway has the most traction, especially amongst the CF population. The vagal pathway is not as well supported but includes stimulation of organs innervated by the vagal nerve resulting in peripheral vasodilation of the extremities [84]. However, to date, neither mechanism has been unequivocally supported with evidence.

2. Conclusion

Musculoskeletal manifestations are common in Cystic Fibrosis patients with most patients having arthropathy, muscular dysfunction or decreased bone quality at some point during their lifetime. The decrease in BMD in CF patients leads to an elevated risk of both appendicular and axial skeletal fractures, which can be mitigated with allele specific drugs due to their ability to return osteoblast function to near normal levels, while also significantly improving BMD. Vitamin D supplementation is also an important adjunct to maximize both bone health and muscular function. Although a combination of factors ultimately leads to skeletal and muscular dysfunction, targeted exercise and resistance programs have been shown to be effective at improving both BMD and muscular function (**Figure 4**).

As CF patients have improved life expectancies due to rapid improvements in pharmaceuticals and CF treatment protocols, the prevalence of arthropathy will continue to increase. Differentiating CFA and HOA from septic arthritis via non-invasive measures can help minimize unnecessary procedures including arthrocentesis, while optimizing outcomes. Additionally, both CFA and HOA can be treated with NSAIDs

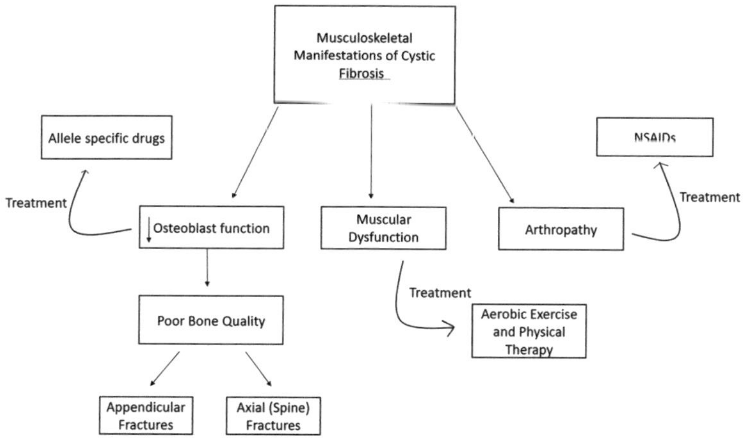

Figure 4.
Treatment algorithm for the varying presentations of the musculoskeletal manifestations of Cystic Fibrosis.

with abrupt minimization of symptoms. Future research is necessary to document the role of allele specific drugs in improving the musculoskeletal manifestations of Cystic Fibrosis.

Conflict of interest

The author declares no conflict of interest.

Author details

Mark Lambrechts
Rothman Orthopaedic Institute at Thomas Jefferson University, Philadelphia, PA, USA

*Address all correspondence to: mark.lambrechts2016@gmail.com

References

[1] Sala MA, Jain M. Tezacaftor for the treatment of cystic fibrosis. Expert Review of Respiratory Medicine. 2018;**12**(9):725-732

[2] Rowe SM, Heltshe SL, Gonska T, et al. Clinical mechanism of the cystic fibrosis transmembrane conductance regulator potentiator ivacaftor in G551D-mediated cystic fibrosis. American Journal of Respiratory and Critical Care Medicine. 2014;**190**(2):175-184

[3] Scotet V, L'Hostis C, Férec C. The changing epidemiology of cystic fibrosis: Incidence, survival and impact of the *CFTR* gene discovery. Genes (Basel). 2020;**11**(6):589

[4] Brunetti G, D'Amato G, Chiarito M, Tullo A, Colaianni G, Colucci S, et al. An update on the role of RANKL-RANK/osteoprotegerin and WNT-ß-catenin signaling pathways in pediatric diseases. World Journal of Pediatrics. 2019;**15**(1):4-11

[5] Turner CH. On Wolff's law of trabecular architecture. Journal of Biomechanics. 1992;**25**(1):1-9

[6] Courtney JM, Ennis M, Elborn JS. Cytokines and inflammatory mediators in cystic fibrosis. Journal of Cystic Fibrosis. 2004;**3**(4):223-231

[7] Shead EF, Haworth CS, Barker H, Bilton D, Compston JE. Osteoclast function, bone turnover and inflammatory cytokines during infective exacerbations of cystic fibrosis. Journal of Cystic Fibrosis. 2010;**9**(2):93-98

[8] Sermet-Gaudelus I, Delion M, Durieu I, Jacquot J, Hubert D. Bone demineralization is improved by ivacaftor in patients with cystic fibrosis carrying the p.Gly551Asp mutation. Journal of Cystic Fibrosis. 2016;**15**(6):e67-e69

[9] Delion M, Braux J, Jourdain ML, Guillaume C, Bour C, Gangloff S, et al. Overexpression of RANKL in osteoblasts: A possible mechanism of susceptibility to bone disease in cystic fibrosis. The Journal of Pathology. 2016;**240**(1):50-60

[10] Ambroszkiewicz J, Sands D, Gajewska J, Chelchowska M, Laskowska-Klita T. Bone turnover markers, osteoprotegerin and RANKL cytokines in children with cystic fibrosis. Advances in Medical Sciences. 2013;**58**:338-343

[11] Le Henaff C, Haÿ E, Velard F, et al. Enhanced F508del-CFTR channel activity ameliorates bone pathology in murine cystic fibrosis. The American Journal of Pathology. 2014;**184**(4):1132-1141

[12] Haworth CS, Webb AK, Egan JJ, et al. Bone histomorphometry in adult patients with cystic fibrosis. Chest. 2000;**118**(2):434-439

[13] Velard F, Delion M, Le Henaff C, et al. Cystic fibrosis and bone disease: Defective osteoblast maturation with the F508del mutation in cystic fibrosis transmembrane conductance regulator. American Journal of Respiratory and Critical Care Medicine. 2014;**189**(6):746-748

[14] Velard F, Delion M, Lemaire F, et al. Cystic fibrosis bone disease: Is the CFTR corrector C18 an option for therapy? The European Respiratory Journal. 2015;**45**(3):845-848

[15] Dif F, Marty C, Baudoin C, de Vernejoul MC, Levi G. Severe

osteopenia in CFTR-null mice. Bone. 2004;**35**(3):595-603

[16] Brookes DS, Briody JN, Munns CF, Davies PS, Hill RJ. Cystic fibrosis-related bone disease in children: Examination of peripheral quantitative computed tomography (pQCT) data. Journal of Cystic Fibrosis. 2015;**14**(5):668-677

[17] Putman MS, Milliren CE, Derrico N, et al. Compromised bone microarchitecture and estimated bone strength in young adults with cystic fibrosis. The Journal of Clinical Endocrinology and Metabolism. 2014;**99**(9):3399-3407

[18] Sermet-Gaudelus I, Souberbielle JC, Ruiz JC, Vrielynck S, Heuillon B, Azhar I, et al. Low bone mineral density in young children with cystic fibrosis. American Journal of Respiratory and Critical Care Medicine. 2007;**175**:951-957

[19] Hind K, Truscott JG, Conway SP. Exercise during childhood and adolescence: A prophylaxis against cystic fibrosis-related low bone mineral density? Exercise for bone health in children with cystic fibrosis. Journal of Cystic Fibrosis. 2008;**7**(4):270-276. DOI: 10.1016/j.jcf.2008.02.001

[20] Pashuck TD, Franz SE, Altman MK, Wasserfall CH, Atkinson MA, Wronski TJ, et al. Murine model for cystic fibrosis bone disease demonstrates osteopenia and sex-related differences in bone formation. Pediatric Research. 2009;**65**(3):311-316

[21] King SJ, Topliss DJ, Kotsimbos T, Nyulasi IB, Bailey M, Ebeling PR, et al. Reduced bone density in cystic fibrosis: DeltaF508 mutation is an independent risk factor. The European Respiratory Journal. 2005;**25**(1):54-61

[22] Legroux-Gérot I, Leroy S, Prudhomme C, Perez T, Flipo RM, Wallaert B, et al. Bone loss in adults with cystic fibrosis: Prevalence, associated factors, and usefulness of biological markers. Joint, Bone, Spine. 2012;**79**(1):73-77

[23] Rana M, Munns CF, Selvadurai H, Briody J, Craig ME. The impact of dysglycaemia on bone mineral accrual in young people with cystic fibrosis. Clinical Endocrinology. 2013;**78**(1): 36-42

[24] Gronowitz E, Mellström D, Strandvik B. Normal annual increase of bone mineral density during two years in patients with cystic fibrosis. Pediatrics. 2004;**114**(2):435-442

[25] Hardin DS, Arumugam R, Seilheimer DK, LeBlanc A, Ellis KJ. Normal bone mineral density in cystic fibrosis. Archives of Disease in Childhood. 2001;**84**(4):363-368

[26] Aris RM, Merkel PA, Bachrach LK, et al. Guide to bone health and disease in cystic fibrosis. The Journal of Clinical Endocrinology and Metabolism. 2005;**90**(3):1888-1896

[27] Khundmiri SJ, Murray RD, Lederer E. PTH and vitamin D. Comprehensive Physiology. 2016;**6**(2):561-601

[28] Paccou J, Zeboulon N, Combescure C, Gossec L, Cortet B. The prevalence of osteoporosis, osteopenia, and fractures among adults with cystic fibrosis: A systematic literature review with meta-analysis. Calcified Tissue International. 2010;**86**(1):1-7

[29] McCauley LA, Thomas W, Laguna TA, Regelmann WE, Moran A, Polgreen LE. Vitamin D deficiency is associated with pulmonary exacerbations in children with cystic fibrosis. Annals of the American Thoracic Society. 2014;**11**(2):198-204

[30] Lark RK, Lester GE, Ontjes DA, Blackwood AD, Hollis BW, Hensler MM, et al. Diminished and erratic absorption of ergocalciferol in adult cystic fibrosis patients. The American Journal of Clinical Nutrition. 2001;**73**(3):602-606

[31] Tangpricha V, Lukemire J, Chen Y, et al. Vitamin D for the immune system in cystic fibrosis (DISC): A double-blind, multicenter, randomized, placebo-controlled clinical trial. The American Journal of Clinical Nutrition. 2019;**109**(3):544-553

[32] Sermet-Gaudelus I, Bianchi ML, Garabédian M, et al. European cystic fibrosis bone mineralisation guidelines. Journal of Cystic Fibrosis. 2011;**10** (Suppl 2):S16-S23

[33] Chapman I, Greville H, Ebeling PR, King SJ, Kotsimbos T, Nugent P, et al. Intravenous zoledronate improves bone density in adults with cystic fibrosis (CF). Clinical Endocrinology. 2009;**70**(6):838-846

[34] Haworth CS, Selby PL, Adams JE, Mawer EB, Horrocks AW, Webb AK. Effect of intravenous pamidronate on bone mineral density in adults with cystic fibrosis. Thorax. 2001;**56**(4):314-316

[35] Aris RM, Lester GE, Caminiti M, Blackwood AD, Hensler M, Lark RK, et al. Efficacy of alendronate in adults with cystic fibrosis with low bone density. American Journal of Respiratory and Critical Care Medicine. 2004;**169**(1):77-82

[36] Conwell LS, Chang AB. Bisphosphonates for osteoporosis in people with cystic fibrosis. Cochrane Database of Systematic Reviews. 2014;**2014**(3):CD002010. DOI: 10.1002/14651858.CD002010.pub4

[37] Compston JE, McClung MR, Leslie WD. Osteoporosis. Lancet. 2019;**393**(10169):364-376

[38] Kates SL, Ackert-Bicknell CL. How do bisphosphonates affect fracture healing? Injury. 2016;**47**(Suppl 1(0 1)): S65-S68

[39] Xue D, Li F, Chen G, Yan S, Pan Z. Do bisphosphonates affect bone healing? A meta-analysis of randomized controlled trials. Journal of Orthopaedic Surgery and Research. 2014;**5**(9):45

[40] Gao Y, Liu X, Gu Y, et al. The effect of bisphosphonates on fracture healing time and changes in bone mass density: A meta-analysis. Front Endocrinology (Lausanne). 2021;**12**:688269 Published 2021 Aug 30

[41] Ju DG, Mogayzel PJ Jr, Sponseller PD, Familiari F, McFarland EG. Bilateral midshaft femoral fractures in an adolescent baseball player. Journal of Cystic Fibrosis. 2016;**15**(4):e41-e43

[42] Lim AY, Isopescu S, Thickett KM, et al. Bilateral fractured neck of the femur in an adult patient with cystic fibrosis. European Journal of Internal Medicine. 2003;**14**(3):196-198

[43] Haworth CS, Freemont AJ, Webb AK, et al. Hip fracture and bone histomorphometry in a young adult with cystic fibrosis. The European Respiratory Journal. 1999;**14**(2):478-479

[44] Al-Azzani WA, Evans L, Speight L, et al. Hyperpharmacotherapy in ageing cystic fibrosis patients: The first report of an atypical hip fracture. Respiratory Medicine Case Reports. 2015;**16**:148-150

[45] Ross J, Gamble J, Schultz A, Lewiston N. Back pain and spinal deformity in cystic fibrosis. American Journal of Diseases of Children. 1987;**141**(12):1313-1316

[46] Lambrechts MJ, Smith MJ, Choma TJ. Orthopedic manifestations

of cystic fibrosis. Orthopedics. 2021;**44**(3):e440-e445. DOI: 10.3928/01477447-20210415-03

[47] Kumar N, Balachandran S, Millner PA, Littlewood JM, Conway SP, Dickson RA. Scoliosis in cystic fibrosis: Is it idiopathic? Spine. 2004;**29**(18):1990-1995

[48] Tejero S, Cejudo P, Quintana-Gallego E, Sañudo B, Oliva-Pascual-Vaca A. The role of daily physical activity and nutritional status on bone turnover in cystic fibrosis: A cross-sectional study. Brazilian Journal of Physical Therapy. 2016;**20**(3):206-212

[49] Aris RM, Renner JB, Winders AD, et al. Increased rate of fractures and severe kyphosis: Sequelae of living into adulthood with cystic fibrosis. Annals of Internal Medicine. 1998;**128**(3):186-193

[50] de Meer K, Gulmans VA, van Der Laag J. Peripheral muscle weakness and exercise capacity in children with cystic fibrosis. American Journal of Respiratory and Critical Care Medicine. 1999;**159**(3):748-754

[51] Hathorn C, Fall A, McGurk S, Tsirikos AI, Urquhart DS. G101(P) incidence of scoliosis in adolescent cystic fibrosis patients. Archives of Disease in Childhood. 2014;**99**(suppl 1):A43

[52] Weinstein SL, Ponseti IV. Curve progression in idiopathic scoliosis. The Journal of Bone and Joint Surgery. American Volume. 1983;**65**(4):447-455

[53] Troosters T, Langer D, Vrijsen B, et al. Skeletal muscle weakness, exercise tolerance and physical activity in adults with cystic fibrosis. The European Respiratory Journal. 2009;**33**(1):99-106

[54] Divangahi M, Balghi H, Danialou G, et al. Lack of CFTR in skeletal muscle predisposes to muscle wasting and diaphragm muscle pump failure in cystic fibrosis mice. PLoS Genetics. 2009;**5**(7):e1000586

[55] Rosenthal M, Narang I, Edwards L, Bush A. Non-invasive assessment of exercise performance in children with cystic fibrosis (CF) and non-cystic fibrosis bronchiectasis: Is there a CF specific muscle defect? Pediatric Pulmonology. 2009;**44**(3):222-230

[56] King SJ, Nyulasi IB, Bailey M, Kotsimbos T, Wilson JW. Loss of fat-free mass over four years in adult cystic fibrosis is associated with high serum interleukin-6 levels but not tumour necrosis factor-alpha. Clinical Nutrition. 2014;**33**(1):150-155

[57] Lamhonwah AM, Bear CE, Huan LJ, Kim Chiaw P, Ackerley CA, Tein I. Cystic fibrosis transmembrane conductance regulator in human muscle: Dysfunction causes abnormal metabolic recovery in exercise. Annals of Neurology. 2010;**67**(6):802-808

[58] Gruet M, Decorte N, Mely L, Vallier JM, Camara B, Quetant S, et al. Skeletal muscle contractility and fatigability in adults with cystic fibrosis. Journal of Cystic Fibrosis. 2016;**15**(1):e1-e8

[59] Shah AR, Gozal D, Keens TG. Determinants of aerobic and anaerobic exercise performance in cystic fibrosis. American Journal of Respiratory and Critical Care Medicine. 1998;**157**(4 Pt 1):1145-1150

[60] Boas SR, Joswiak ML, Nixon PA, Fulton JA, Orenstein DM. Factors limiting anaerobic performance in adolescent males with cystic fibrosis. Medicine and Science in Sports and Exercise. 1996;**28**(3):291-298

[61] Gilsanz V, Kremer A, Mo AO, Wren TA, Kremer R. Vitamin D status and its relation to muscle mass and muscle fat in young women. The Journal of Clinical Endocrinology and Metabolism. 2010;**95**(4):1595-1601

[62] King SJ, Tierney AC, Edgeworth D, et al. Body composition and weight changes after ivacaftor treatment in adults with cystic fibrosis carrying the G551 D cystic fibrosis transmembrane conductance regulator mutation: A double-blind, placebo-controlled, randomized, crossover study with open-label extension. Nutrition. 2021;**85**:111124

[63] Stallings VA, Sainath N, Oberle M, Bertolaso C, Schall JI. Energy balance and mechanisms of weight gain with Ivacaftor treatment of cystic fibrosis gating mutations. The Journal of Pediatrics. 2018;**201**:229-237.e4

[64] Gruet M, Troosters T, Verges S. Peripheral muscle abnormalities in cystic fibrosis: Etiology, clinical implications and response to therapeutic interventions. Journal of Cystic Fibrosis. 2017;**16**(5):538-552

[65] Rovedder PM, Flores J, Ziegler B, Casarotto F, Jaques P, Barreto SS, et al. Exercise programme in patients with cystic fibrosis: A randomized controlled trial. Respiratory Medicine. 2014;**108**(8):1134-1140

[66] Strauss GD, Osher A, Wang CI, Goodrich E, Gold F, Colman W, et al. Variable weight training in cystic fibrosis. Chest. 1987;**92**(2):273-276

[67] Bieli C, Summermatter S, Boutellier U, Moeller A. Respiratory muscle training improves respiratory muscle endurance but not exercise tolerance in children with cystic fibrosis. Pediatric Pulmonology. 2017;**52**(3): 331-336. DOI: 10.1002/ppul.23647

[68] Santana-Sosa E, Gonzalez-Saiz L, Groeneveld IF, Villa-Asensi JR, Gómez B, de Aguero MI, et al. Benefits of combining inspiratory muscle with 'whole muscle' training in children with cystic fibrosis: A randomised controlled trial. British Journal of Sports Medicine. 2014;**48**(20):1513-1517

[69] Swisher AK, Hebestreit H, Mejia-Downs A, et al. Exercise and habitual physical activity for people with cystic fibrosis. Cardiopulmonary Physical Therapy Journal. 2015;**26**(4):85-98

[70] Botton E, Saraux A, Laselve H, Jousse S, Le Goff P. Musculoskeletal manifestations in cystic fibrosis. Joint, Bone, Spine. 2003;**70**(5):327-335

[71] Grehn C, Dittrich AM, Wosniok J, Holz F, Hafkemeyer S, Naehrlich L, et al. Registry working group of the German CF registry. Risk factors for cystic fibrosis arthropathy: Data from the German cystic fibrosis registry. Journal of Cystic Fibrosis. 2021;**20**(6):e87-e92

[72] Roehmel JF, Kallinich T, Staab D, Schwarz C. Clinical manifestations and risk factors of arthropathy in cystic fibrosis. Respiratory Medicine. 2019;**147**:66-71

[73] Nixon LS, Yung B, Bell SC, Elborn JS, Shale DJ. Circulating immunoreactive interleukin-6 in cystic fibrosis. American Journal of Respiratory and Critical Care Medicine. 1998;**157**(6 pt 1):1764-1769

[74] Stannus O, Jones G, Cicuttini F, et al. Circulating levels of IL-6 and TNF-alpha are associated with knee radiographic osteoarthritis and knee cartilage loss in older adults. Osteoarthritis and Cartilage. 2010;**18**(11):1441-1447

[75] Lands LC, Stanojevic S. Oral non-steroidal anti-inflammatory drug therapy for lung disease in cystic fibrosis.

Cochrane Database of Systematic Reviews. 2019;**9**:CD001505

[76] Loh G, Ryaboy I, Skabelund A, French A. Procalcitonin, erythrocyte sedimentation rate and C-reactive protein in acute pulmonary exacerbations of cystic fibrosis. The Clinical Respiratory Journal. 2018;**12**(4):1545-1549

[77] Matouk E, Nguyen D, Benedetti A, Bernier J, Gruber J, Landry J, et al. C-reactive protein in stable cystic fibrosis: An additional indicator of clinical disease activity and risk of future pulmonary exacerbations. Journal of Pulmonary Respiratory Medicine. 2016;**6**(5):1000375

[78] Talebi-Taher M, Shirani F, Nikanjam N, Shekarabi M. Septic versus inflammatory arthritis: Discriminating the ability of serum inflammatory markers. Rheumatology International. 2013;**33**(2):319-324

[79] Turner MA, Baildam E, Patel L, David TJ. Joint disorders in cystic fibrosis. Journal of the Royal Society of Medicine. 1997;**90**(Suppl 31):13-20

[80] Johnson S, Knox AJ. Arthropathy in cystic fibrosis. Respiratory Medicine. 1994;**88**(8):567-570

[81] Prajapati BB, Filippi A, Sears EH. Chronic joint pain in a young adult with cystic fibrosis. Cureus. 2021;**13**(8):e17229 [Published 2021 Aug 16]

[82] Tekiteki A, Good WR, Diggins B, Anderson G, Wong CA. Recurrent hypertrophic pulmonary osteoarthropathy in an adult with bronchiectasis. Respirology Case Reports. 2020;**8**(6):e00602 [Published 2020 Jun 23]

[83] Yap FY, Skalski MR, Patel DB, Schein AJ, White EA, Tomasian A, et al. Hypertrophic Osteoarthropathy: Clinical and imaging features. Radiographics. 2017;**37**(1):157-195

[84] Treasure T. Hypertrophic pulmonary osteoarthropathy and the vagus nerve: An historical note. Journal of the Royal Society of Medicine. 2006;**99**(8):388-390

Section 3

Recent Developments in the Treatment of Skeletal Muscle Disorders

Chapter 4

Duchenne Muscular Dystrophy: Clinical and Therapeutic Approach

Radenka Kuzmanić Šamija and Marta Plejić

Abstract

Duchenne muscular dystrophy (DMD) and Becker muscular dystrophy (BMD) are dystrophinopathies, a group of muscular dystrophies caused by mutations in the dystrophin gene. Duchenne muscular dystrophy is the most common muscular dystrophy that occurs in children. A mutation in the DMD gene leads to a loss of expression of the dystrophin protein, a subsarcolemmal protein that provides strength, stability, and functionality to the myofibrils. Patients with dystrophinopathies with basic progressive weakness of the musculoskeletal system develop complications of many organ systems that significantly contribute to the deterioration of the clinical condition and shorter life expectancy. Multidisciplinary care has extended the patients' life expectancy and the development of subspecialist branches has enabled the improvement of diagnostic methods and treatment. Recently, therapeutic options in the treatment of DMD have advanced significantly, and new genetic and molecular therapies are emerging. The advent of gene therapy as a causal therapy for DMD has placed additional emphasis on diagnosing and treating the disease as early as possible. This achieves an additional prolongation of life expectancy, increases the quality of life in patients with DMD, and provides hope for patients and their families.

Keywords: muscular dystrophy, skeletal muscles, children, Duchenne, genetic therapy, ataluren

1. Introduction

Muscular dystrophies are inherited, progressive muscle disorders resulting from defects in one or more genes needed for normal muscle structure and function. Muscular dystrophies are distinguished by the selective distribution of weakness and the specific nature of the genetic abnormality involved, which affect all races and ethnic groups. Duchenne muscular dystrophy (DMD) and Becker muscular dystrophy (BMD) are dystrophinopathies, a group of muscular dystrophies caused by mutations in the dystrophin gene. DMD is a severe X-linked recessive neuromuscular disorder caused by mutations in the dystrophin gene that result in absent or insufficient functional dystrophin, a cytoskeletal protein that enables the strength, stability, and functionality of myofibers. Dystrophin is bound to actin in the cytosol by a transmembrane complex and connects the cytoskeleton and extracellular matrix, allowing contractility of the muscle fiber. Gene that encodes dystrophin is one of the largest genes in the human body. It occupies about 1% of the whole x-chromosome. Duchenne muscular dystrophy is the most common muscular dystrophy that occurs in children, with a frequency of 1 in 3500 live-born male children. In contrast, Becker muscular dystrophy occurs much less frequently, approximately 1 in 30,000 live-born male children. According to epidemiological data, frequency occurrence is approximately the same in all countries of the world [1].

2. Pathophysiology

DMD is inherited in an X-linked pattern. The gene that can carry a DMD-causing mutation is on the X chromosome. Every boy inherits an X chromosome from his mother and a Y chromosome from his father. Girls get two X chromosomes, one from mother and one from father. Due to the X-linked inheritance, each son born to a woman with a dystrophin mutation on one of her two X chromosomes has a 50 percent chance of inheriting the flawed gene and having DMD. Each of her daughters has a 50 percent chance of inheriting the mutation and being a carrier who may pass the mutation onto their own children. Carriers may not have any disease symptoms but can have a child with the mutation or the disease [1]. However, the latest research shows that in some cases, female carriers may develop symptoms such as cardiomyopathy. It is important to note that one-third of boys with DMD are born to mothers without mutation in the X chromosome due to a new genetic mutation that arose in one of mother's egg cells. Mutations in the DMD gene that cause phenotypic expression of the disease are numerous and highly variable. They include deletions of the whole gene, deletions or duplications of one or more exons, and insertions or alterations of a single nucleotide. Deletions of one or more exons of the DMD gene represent 60–70% of mutations in Duchenne muscular dystrophy, and 80–90% of mutations in Becker muscular dystrophies. Point mutations make up 25–35% of Duchenne mutations and 10–20% mutations in Becker muscular dystrophy. Duplications are the least represented, with a frequency of 5–10% in Duchenne and Becker muscular dystrophy. In more than 90% of patients, dystrophin deficiency is the result of "out of frame" mutations that cause the reading frame to shift. Such mutations cause transcription termination of informational RNA, resulting in complete dystrophin deficiency or the formation of very small amounts of abnormal dystrophin. If "in frame" mutations occur, which do not cause shift of the reading frame, there is a quantitative and qualitative change in dystrophin, and the clinical presentation is less severe. DMD mutations cause virtually no functional dystrophin to be made. On the other hand, patients with BMD make dystrophin that is partially functional, so the clinical presentation is less severe. They make a shortened form of the protein, which protects the muscles in BMD from degenerating as completely or as quickly as those of patients with DMD. Qualitative changes in dystrophin and its deficiency lead to membrane degradation. In the process of cell membrane breakdown, there is also a loss of certain muscle enzymes causing muscle weakness [2].

Previously said, DMD is inherited in an X-linked pattern, which means male children get affected. Female carriers may not have any disease symptoms but can have a child with the mutation or the disease [1]. However, the latest research shows that in some cases, female carriers may develop symptoms such as cardiomyopathy. It is important to note that one-third of boys with DMD are born to mothers without mutation in the X chromosome due to a new genetic mutation that arose in one of mother's egg cells. Mutations in the DMD gene that cause phenotypic expression of the disease are numerous and highly variable. They include deletions of the whole gene, deletions or duplications of one or more exons, insertions, or alterations of a single nucleotide. Deletions of one or more exons of the DMD gene represent 60–70% of mutations in Duchenne muscular dystrophy, and 80–90% of mutations in Becker muscular dystrophies. Point mutations make up 25–35% of Duchenne mutations and 10–20% mutations in Becker muscular dystrophy. Duplications are the least represented, with a frequency of 5–10% in Duchenne and Becker muscular dystrophy. In more than 90% of patients, dystrophin deficiency is the result of "out of frame" mutations that cause the reading frame to shift. Such mutations cause transcription termination of informational RNA, resulting in complete dystrophin deficiency or the formation of very small amounts of abnormal dystrophin. If "in frame" mutations occur, which do not cause shift in the reading frame, there is a quantitative and qualitative change in dystrophin, and the clinical presentation is less severe. Patients with BMD make dystrophin that is partially functional. They make a shortened form of the protein, which protects the muscles in BMD from degenerating as completely or as quickly as those of patients with DMD. Qualitative changes in dystrophin and its deficiency lead to membrane degradation. In the process of cell membrane breakdown, there is also a loss of certain muscle enzymes causing muscle weakness [2].

3. Clinical presentation

In Duchenne muscular dystrophy, no abnormality is noted in the patient at birth, and manifestations of the muscle weakness do not begin until the child begins to walk. Clinical symptoms of Duchenne muscular dystrophy are usually noticed around the age of 3, most often between the ages of 3 and 5. However, taking a more detailed history often reveals that motor development has been slower before. Children with Duchenne may be slower to sit, stand, or walk. Most are unable to run and jump properly due to weakness in the muscles of the body. Self-walking occurs later, approximately at 15 months of age. Signs and symptoms of DMD, which typically appear in early childhood, might include: later onset of independent sitting, frequent falls, clumsiness, difficulty rising from a lying or sitting position, trouble running and jumping, waddling gait, walking on the toes, muscle pain and stiffness, large calf muscles, learning disabilities, and delayed growth. In toddlers, parents may notice enlarged calf muscles. This enlargement is known as pseudohypertrophy, or "false enlargement," because the abnormal muscle tissue is replaced by a fat tissue. A baby or a toddler with DMD may seem clumsy and fall often. Parents may also note that children have trouble climbing stairs, getting up from the floor, or running. When arising from the floor, patients may use hand support to push themselves to an upright position. It is called the Gowers sign. In 1879, a British neurologist, Sir William Richard Gowers, described the most significant Gowers sign as the characteristic pattern seen in patients with Duchenne muscular dystrophy wherein they "climb up" their thighs with the aid of their hands to overcome the weakness of their pelvic and proximal lower limb muscles [1, 2]. Weakness of the hip girdle and upper thigh muscles leads to an instability of the pelvis on standing and walking. If the muscles extending the hip joint are affected, the posture in the hip becomes flexed, and lumbar lordosis increases. Patients with DMD usually have difficulties standing up from a sitting position, so they need to use strength from the arm muscles. Due to weakness in the gluteus medius muscle, the hip on the side of the swinging leg drops with each step, so the waddling gait appears. On average, muscles lose approximately

2% of their strength each year. According to the time period and disease progression, the course of DMD is classified into five phases: diagnosis (infancy/childhood), early ambulatory (childhood), late ambulatory (late childhood/adolescent/young adult), early nonambulatory (adolescent/young adult), and late nonambulatory (adult). Ambulation is the ability to walk without the need for any kind of assistance. Classification is important for establishing diagnostic and treatment protocols that are carried out at a particular stage [2].

Loss of ambulation occurs in untreated patients at the end of the first decades of life, while in patients treated with corticosteroids it occurs two to three years later. Respiratory problems are the main cause of mortality and morbidity in patients with DMD. Respiratory function gradually deteriorates due to weakening of the respiratory muscles, intercostal, and diaphragm. All respiratory functions are affected–oxygen exchange, mucociliary activity, and respiratory control during wakefulness and sleep. Significant cause of mortality and morbidity is cardiac complications. Lack of dystrophin in the heart leads to the development of cardiomyopathy, progressive myocardial fibrosis leading to ventricular dysfunction and sometimes life-threatening cardiac arrhythmias. Body weight can vary from malnutrition to normal values or malnutrition. Glucocorticoid therapy increases appetite and sodium and water retention, and due to muscle weakness, patients have limited physical activity. During adolescence, the risk of malnutrition increases due to dysphagia, mandibular contracture, and constipation. Endocrinological complications that may occur due to glucocorticoid therapy include reduced growth, delayed puberty, and, rarely, adrenal insufficiency. Some patients with DMD may have certain cognitive difficulties, and it is believed that the expression of dystrophin in the brain is variable and that depends on the type of gene mutation. The life expectancy of patients with DMD was approximately 20 years [1, 2].

Advances in diagnostic and therapeutic options have occurred to increase life expectancy by 10 and several years, and it is believed that the gene therapy, which is already underway, will further extend the life expectancy of people with DMD [2].

4. Diagnosis

Achieving a timely and accurate diagnosis of DMD and BMD is a crucial aspect of care. It is extremely important to recognize the early signs of muscular dystrophy, which, unfortunately, are often interpreted as a child's clumsiness or laziness, so they are often not recognized until the 3rd or 4th year of life. Nonmotor manifestations (especially slow speech development) may be failed to observe, neuromuscular diseases are rare, and the attending physician often has no experience with this disease, which contributes to the delay of diagnosis for up to 30 months and the later start of pharmacological therapy and rehabilitation. The diagnosis is most often set in early childhood by the appearance of specific signs, such as muscle weakness, clumsiness, toe walking, difficulty climbing stairs, and a positive Gowers sign. When these symptoms occur, the patient first is referred to a pediatric neurologist who sets suspicion on muscular dystrophy. The role of laboratory tests is extremely important. The most important screening test for dystrophy is determination of serum creatine kinase (CK) levels with determination of alanine aminotransferase (ALT) and aspartate aminotransferase (AST). Slightly elevated or normal CK values are a sufficient condition to rule out the disease. In patients with DMD, CK, ALT, and AST levels are elevated from the early stages of the disease. Unfortunately, there are cases of patients with minimal clinical symptoms at an early age who have missed CK testing, and routine laboratory tests showed elevated ALT and AST values. That situation led them to the wrong focus on liver disease treatment and contributed to delayed diagnosis. In order to diagnose DMD and BMD, it is mandatory to perform genetic testing to determine the exact type and location of the mutation, which also affects the possibility of treatment with specific drugs for certain types of mutations [2].

Almost 70% of patients with DMD have a single-exon or multi-exon deletion or duplication in the dystrophin gene, so that is the reason why the first confirmatory test is dystrophin gene deletion and duplication testing. It is done by multiplex ligation-dependent probe amplification (MLPA) or comparative genomic hybridization array (array CGH) since use of multiplex polymerase chain reaction (PCR) can only identify deletions. Identification of the boundaries of a deletion or duplication mutation by MLPA or comparative genomic hybridization array could indicate whether the mutation is predicted to preserve or disrupt the reading frame. If deletion or duplication testing is negative, genetic sequencing should be done to screen for the types of mutations that are attributed to DMD. These mutations include *point* mutations (nonsense or missense), small deletions, and small duplications or insertions, which can be identified using next-generation sequencing. Finally, if genetic testing does not confirm a clinical diagnosis of DMD, then a muscle biopsy sample should be tested for the presence of dystrophin protein by immunohistochemistry of tissue cryosections or by western blot of a muscle protein extract. In muscle biopsy, the presence of dystrophins is determined by immunohistochemical methods and the amount and size of dystrophin. However, a muscle biopsy can confirm the diagnosis but not the treatment options, as it does not provide information on the type and location of the mutation [1].

5. Treatment

Multidisciplinary approach to care is essential for optimal management of the primary manifestations and for preventing secondary complications of Duchenne muscular dystrophy. Contemporary care has been organized by the availability of more sensitive diagnostic techniques and the earlier use of therapeutic interventions, which have the potential to improve patients' duration and quality of life. The neuromuscular specialist (usually pediatric neurologist) is qualified to guide patients and their families through the increasingly complex and technological diagnostic and therapeutic landscape of contemporary DMD care. Centers for Disease Control and Prevention (CDC) in cooperation with relevant organizations (TREAT-NMD network for neuromuscular diseases, the Muscular Dystrophy Association, and Parent Project Muscular Dystrophy) updated and upgraded recommendations from 2010 in the monitoring and treatment of patients with DMD. Division is essential for establishing a protocol for diagnostics and treatments carried out in each phase [2].

The mainstays of DMD remain physiotherapy and glucocorticoid treatment, which should be continued after the stage of loss of ambulation. The benefits of long-term glucocorticoid therapy have been shown to include loss of ambulation at a later age, preserved upper limb and respiratory function, and avoidance of scoliosis surgery. Recent studies confirm the benefits of starting glucocorticoids in younger children, before significant physical decline. Although the benefits of glucocorticoid therapy are well established, uncertainty remains fact about which glucocorticoids are best and at what doses. Glucocorticoids used in the treatment of DMD and BMD are prednisone and deflazacort (an oxazolone derivative of prednisone).

As a starting dose, it is recommended to take prednisone at a dose of 0.75 mg/kg or deflazacort at a dose of 0.9 mg/kg [2]. The main side effects of glucocorticoids are decreased bone density, increased appetite and body weight, and growth retardation. In studies comparing prednisone and deflazacort, a smaller one was observed weight gain of patients on deflazacort therapy. It is important to note that the FDA approved deflazacort, making it the first glucocorticoid with a specific indication in the treatment of DMD. In cases of intolerance to side effects, recommendations are a reduction in the dose of glucocorticoids by 25–33%, with a revision of symptoms for a month [2, 3].

DMD is characterized by progressive muscle degeneration and weakness, risk of progressive contracture and deformity, decrease in pulmonary and heart function, and functional losses resulting from dystrophin deficiency. Improved DMD management has resulted in prolongation of ambulation, decreased prevalence of severe contracture and deformity, and prolonged function in all areas of life. Rehabilitation team includes physicians, physical therapists, occupational therapists, speech pathologists, and medical equipment providers. Rehabilitation management requires an understanding of DMD pathology, and disease progression [3, 4].

Rehabilitation assessment includes measures of passive ranges of motion, muscle extensibility, posture and alignment, strength, and function in all activities of everyday life. Specialized functional assessment includes analysis of patterns of movement and standardized assessments specific to DMD and other neuromuscular disorders. Foundational clinical assessments of function during the ambulatory period are the North Star Ambulatory Assessment (NSAA) and timed function tests and should be done every 6 months. The NSAA and timed function tests have high reliability and validity, as well as a correlation between tests across time and predictive capabilities regarding functional motor changes that are important in monitoring clinical progression and assessing new and emerging therapies. Identification of optimally responsive test ranges improves predictive capabilities. The North Star Ambulatory Assessment (NSAA) is a 17-item rating scale that is used to measure functional motor abilities in ambulant patients with DMD. It is usually used to monitor the progression of the disease and treatment effects. The NSAA scale consists of 17 components that are scored depending on the patient's performance, with 2, 1, or 0 points. The maximum number of points is 34. The components that are assessed in patients are: standing, walking, getting up from a chair, standing on one leg—left and right, climbing stairs—left and right foot, descending the stairs—left and right foot, sitting from a lying position, raising from floor, lifting head, standing on heels, jumping, and hop on right and left leg and running 10 m. The examination must be performed without the use of those. The 6-min walk test is a sub-maximal exercise test used for evaluation of aerobic capacity and endurance. The distance covered over a time of 6 min is used as the outcome by which to compare changes in performance capacity. Tests that predict potential upcoming changes can be used to guide proactive care, such as impairment-level interventions and possible equipment needs. Specifically, before age of 7 years, gains might occur in the 6-min walk test and timed function tests. After 7 years, a 6-min walk test result of less than 325 m, time to stand more than 30 sec, time to climb four stairs more than 8 s, 10-min walk or run time more than 10–12 s, and mean linearized NSAA 34 or less (raw score of nine) have been associated with greater functional decline in ambulation over the subsequent 12 months. Functional assessment includes the assessment of activities of everyday living and the need for adaptive equipment or possible assistive technology. The NSAA, with revision, can be used to test children's motor function from the age of 3. Hip kinematics during gait are clinically meaningful

DOI: http://dx.doi.org/10.5772/intechopen.106017

outcome measures at 4–8 years. Other measures assessing antigravity function include the Alberta Infant Motor Scale, Hammersmith Functional Motor Scale Expanded, and the Gross Motor Function Measure. In older individuals who are nonambulatory. The most popular is The Brooke Upper Extremity Scale, which assesses the function of the upper limb function. Consistent use of the same functional measurements is recommended. Assessment of motor skills by a specialist physiatrist should be monitored every 4-6 months, with more frequent checkups as needed [2, 3].

Direct physical, occupational, and speech therapy should be provided in outpatient and school settings and continue throughout adulthood. The goal of muscle extensibility and joint mobility management is prevention or minimization of the occurrence of contracture and deformities. The inability to move a joint through its full range of motion, muscle imbalance in a joint, chronic static positioning, and fibrotic changes in muscles can cause decreased muscle extensibility and joint contractures. Restricted patterns of breathing and fibrosis of intercostal muscles decrease chest wall mobility. The maintenance of passive ranges of movement, muscle extensibility, and chest wall mobility can optimize movement and functional positioning, maintain ambulation, prevent fixed contractures and deformities, and optimize respiratory function. A daily preventive home stretching program should begin before the loss of passive ranges of motion under the guidance of physical therapist. Regular stretching of ankle, knee, and hip should begin after diagnosis and continue into adulthood. Stretching of the upper extremities is especially important in the stage after loss of ambulation. Power stand-and-drive motorized wheelchairs are now frequently used instead of knee-ankle-foot orthoses to support standing mobility. Such orthoses might still be an appropriate choice in some situations but should be viewed as therapeutic rather than functional tools. Possible adaptive equipment and home renovations include patient lifts for safe transfers, stair lifts, bathing and bathroom equipment or renovations, special beds and mattresses, and vehicle modifications. Physical therapists prescribe, monitor, and guide exercise, which can prevent an unnecessarily sedentary lifestyle and the associated problems of social isolation and overweight. However, the effects of exercise on muscle degeneration in dystrophinopathies can include damage due to structural fragility of muscles, metabolic abnormalities, and reduced exercise capacity. Eccentric muscle activity or exercise and high-resistance exercise or strength training should be avoided. It is recommended to perform the submaximal aerobic exercise. Swimming is highly recommended from the early ambulatory stage and can be frequently continued into adulthood. Warm water allows children with DMD to perform targeted stretches, exercises, and function-based and play activities that are progressively lost to them on dry land. The Halliwick method is very popular. It is the concept for teaching people with any physical, mental, or sensory difficulties. One of the founders of the method, James McMillan, combined the principles of water, hydrostatics and hydrodynamics, with form and behavior bodies in water. The goal is to achieve: improving breathing control, rhythmic movement coordination, sensory integration, stability and mobility control, improving general fitness and health, and social interaction [5].

Patients and their families are at increased risk of depression and anxiety, particularly at major care transition points in the progression of the disease. The neuromuscular care team should include a mental health clinician (psychologist, psychiatrist) who has specialized training and experience in assisting families and patients with chronic medical or neurodevelopmental conditions. DMD may affect a patient's ability to consistently access his educational environment. Accommodations should be provided to maximize a patient's ability to function normally with his peers [5].

The aim of musculoskeletal care is to maintain motor function for as long as possible, minimize joint contractures, promote bone health, and maintain a straight spine. Assessment for scoliosis should be done at least annually. In ambulatory stage, visual assessment is appropriate, with radiographic assessment only if a curve is observed on examination or if visual inspection alone is inadequate, for example, in children with obesity. Inspection of the spine should be done at every clinical examination in an early nonambulatory stage. Experienced clinicians should be able to monitor the spine in nonambulatory boys by inspection, while less experienced clinicians should obtain a spine radiograph when a child first becomes nonambulatory. Once a curve has been detected with radiography, further surveillance depends on the skeletal maturity of the individual. Skeletally immature individuals should undergo radiographs once every 6 months, while skeletally mature individuals should undergo radiographs at least once a year. A curve of 20° or more should warrant involvement of an orthopedic surgeon. It is not recommended to use the spinal orthoses. Patients treated with corticosteroids have milder spinal curvatures and less frequent need for spinal surgeries. In late nonambulatory stage, clinicians should examine the spine at every clinical visit. Individuals with known scoliosis should have yearly anteroposterior upright spinal radiographs when there is any concern about progression [3].

Patients on glucocorticoid therapy as part of DMD treatment often develop osteoporosis. It is often manifested as trauma to the lower spine or fracture of the long bones [4]. This outcome is not surprising because of known toxicity of glucocorticoid therapy on bone density and, in combination with progressive myopathy, the risk for reduction in bone strength. Approximately 20–60% of children with DMD develop long bone fractures (most commonly the distal femur, tibia, and fibula), and about 30% of them have vertebral fractures. Vertebral fractures are often asymptomatic and are mostly found during control radiological images of the spine, so the prevalence is probably higher than available in the literature. If left untreated, vertebral fractures lead to chronic back pain and spinal deformities [4].

Despite the high incidence of fractures, no studies have yet been published on DMD or diseases associated with osteoporosis in which the benefit and safety of the therapy being assessed used in first-line fracture prevention. According to the latest guidelines, it is considered how bone mineral density (BMD) obtained by densitometry is no longer a major factor in assessing bone fragility. Bone mineral density is now supplementing the approach used to identify the earliest signs of impending bone fractures. Due to a large number of vertebral fractures that are asymptomatic, it is important to emphasize how spinal radiographs should be done regularly, regardless of the absence of symptoms [6].

Fractures are more likely to occur in children whose BDM Z value is higher than −2 SD (standard deviation), thus emphasizing the inadequacy of densitometry as the only method for assessing fracture risk. Regarding bone health, the approach of a patient with DMD at diagnosis includes paying attention to the existence of back pain, before starting glucocorticoid therapy, patient should test levels of serum calcium, phosphate, magnesium, alkaline phosphatase, and parathyroid hormone levels. Also, when making a diagnosis once a year, calcium and vitamin D intake should be determined, serum 25-hydroxyvitamin D3 and perform densitometry [7].

Intravenous bisphosphonates are used for treatment of osteoporosis caused by glucocorticoids. Treatment also includes calcium and vitamin D supplementation. Bisphosphonates are synthetic analogs of pyrophosphate that bind to hydroxyapatite in bone and strongly inhibit bone resorption by slowing down maturation and osteoclast activity. Indications for initiating intravenous treatment with bisphosphonates have not changed from before and include the presence of vertebral or long vertebral fracture bones. Renal function should be measured before initiating intravenous bisphosphonate therapy. It is important to emphasize that intravenous bisphosphonates are used in treatment, not oral ones. It is essential to measure the dose over long periods of time, as well as monitor safety and efficacy of treatment [6, 7].

The most common cause of morbidity and mortality in people with DMD is respiratory complications. The most common complications are respiratory muscle fatigue, atelectasis, mucus plugging, pneumonia, and respiratory failure. Untreated complications can put patients at risk of severe dyspnoea, lengthy hospital admissions due to pneumonia, and death due to respiratory arrest or respiratory-induced cardiac arrhythmias [3]. Respiratory management includes monitoring of respiratory muscle function and the timely use of lung volume recruitment, methods of assisted coughing, and nocturnally assisted ventilation and subsequent daytime ventilation. These therapies can decrease respiratory complications, improve quality of life, and prolong patients' survival. Patients should typically be using most of these therapies by the age of 18–21 years [8]. Serial monitoring of pulmonary function is critical for respiratory care. Forced vital capacity (FVC) rises with growth until an individual becomes nonambulatory. FVC reaches a peak, followed by a plateau, and then deteriorates over the time. Deteriorating FVC may occur in the absence of dyspnoea and remain unrecognized unless pulmonary function is measured regularly. Because the rate of change in FVC over time can vary greatly among individuals, serial measurement of FVC is necessary to characterize each individual's respiratory phenotype. During the ambulatory stage, sleep studies with capnography might be necessary, especially for individuals with weight gain due to glucocorticoid therapy and for individuals with symptoms of sleep-disordered breathing. According to guidelines from the US Centers for Disease Control and Prevention, it is important for patients with DMD to receive yearly immunization with the inactivated influenza vaccine (the injectable vaccine, not the live, attenuated nasal vaccine) and pneumococcal vaccines (including PCV13 and PPSV23) Patients and their caregivers should be educated about respiratory complications during the ambulatory stage of DMD to prepare them for possible future medical complications and therapies. The need for respiratory interventions occurs usually after the loss of ambulation. It is recommended to measure FVC, maximum inspiratory and expiratory pressures (MIP and MEP), peak cough flow, and blood oxygen saturation by pulse oximetry (SpO2) at least every 6 months in all nonambulatory individuals. End-tidal or transcutaneous partial pressure of carbon dioxide in the blood should be measured every 6 months or any time SpO2 is 95% or lower in room air. As their vital capacity decreases, patients with DMD develop noncompliant chest walls and lung volume restriction. To preserve lung compliance, lung volume recruitment is indicated when FVC is 60% predicted or less, achieved with a self-inflating manual ventilation bag or mechanical insufflation-exsufflation device to provide deep lung inflation once or twice daily [3, 8].

During the nonambulatory stage, individuals with DMD often develop weak cough efforts, leading them to risk of atelectasis, pneumonia, ventilation-perfusion mismatch, and progression to respiratory failure, especially during respiratory infections. Treatment includes manual and mechanically assisted coughing, which are indicated when FVC is less than 50% predicted, when peak cough flow is less than 270 L/min, or when maximum expiratory pressure is less than 60 cm H2O. When SpO2 is less than 95% in room air, the frequency of assisted coughing should be increased to prevent and treat mucus plugging, atelectasis, and pneumonia. Antibiotic therapy is recommended during acute respiratory infection when individuals have three of the following five signs of pneumonia: elevated white blood count or C-reactive protein concentration, fever, sputum production, a pulmonary infiltrate on chest radiograph, or hypoxemia or respiratory distress [2, 3, 5].

In the late nonambulatory stage, patients with DMD need assisted ventilation to prolong survival. Signs or symptoms of hypoventilation or sleep-disordered breathing are indications for nocturnally assisted ventilation. Symptoms include fatigue, dyspnoea, morning or continuous headaches, frequent nocturnal awakenings, hypersomnolence, difficulty concentrating, awakenings with dyspnoea, and tachycardia. However, some individuals remain asymptomatic despite the presence of hypoventilation. Nocturnally assisted ventilation should be initiated if a patient's FVC is less than 50% predicted, or when the absolute value of maximum inspiratory pressure is less than 60 cm H2O. Nocturnal ventilation is indicated for patients with abnormal sleep studies, such as overnight oximetry, combination oximetry–capnography, and polysomnography with capnography. Nonambulatory patients with symptoms of sleep-disordered breathing should have sleep studies annually. Because patients with DMD need assisted ventilation to treat hypoventilation, nocturnal noninvasively assisted ventilation (rather than continuous positive airway pressure at a constant level) is first-line therapy for individuals with DMD with obstructive sleep apnea [3, 5].

Continuous noninvasive ventilation methods include mouthpiece or sip ventilation with a portable volume ventilator during the day, changing to nasal ventilation with a bi-level pressure device overnight. Ventilation via tracheostomy or noninvasive ventilation is a controversial question. Some centers use time using the ventilator (16 h/day or more) as an indication for tracheostomy. Clinical experience supports the use of noninvasively assisted ventilation for up to 24 h/day. The decision depends on each individual's preference and clinical course, the skills and usual practices of the individual's clinicians, the standard of care, and the availability of home resources, such as overnight nursing. The use of noninvasive respiratory aids is challenging when individuals with very advanced DMD have acute respiratory illnesses and when they have chronic difficulty swallowing their saliva. Continuous ventilation provides life support, so a backup ventilator and a manual resuscitator should be available in case the primary ventilator malfunctions. The ventilation device and battery should attach to the individual's wheelchair for mobility and quality of life [3, 8].

A major cause of disease-related morbidity and mortality among individuals with DMD is cardiovascular complications. Dystrophin deficiency in the heart muscle causes cardiomyopathy. As the disease progresses, the myocardium fails to meet physiological demands, which leads to heart failure. The failing myocardium can also cause life-threatening arrhythmias. Cardiac management has been challenging because the New York Heart Association (NYHA) classification of heart failure relies on reduced exercise tolerance and patients with DMD have skeletal muscle and cardiac disease combined. The symptoms of heart failure in the nonambulatory individual are frequently overlooked. Early diagnosis and treatment are essential to maximize duration and quality of life. The cardiologist should have clinical expertise in diagnosing and treating heart failure and the cardiomyopathy associated with neuromuscular disease [3]. The baseline cardiac assessment includes cardiac medical history, family history, and a physical examination. Electrocardiogram and noninvasive imaging are advised to establish baseline cardiac function and to screen for underlying anatomical. Echocardiography is recommended until at least age 6–7 years when cardiovascular MRI CMR can usually be done without anesthesia. Patients should have an annual cardiac assessment, until the age of 10 years, including electrocardiogram and noninvasive imaging. After the age of 10 years, asymptomatic patients should have a cardiac examination at least annually because of the increased risk of left ventricular dysfunction. First-line therapy for the treatment of heart disease associated with DMD is angiotensin-converting enzyme (ACE) inhibitors or angiotensin receptor blockers (ARBs). Dosing and ACE inhibitor selection are left to the decision of the cardiologist. Pharmacological therapy should be initiated with the appearance of heart failure symptoms or with the abnormalities such as depressed left ventricular ejection fraction, abnormal chamber dimensions, or the presence of myocardial fibrosis noted on imaging studies. Given the absence of dystrophin-specific targeted cardiac therapies, traditional treatment strategies for heart failure should be used. Progressive myocardial fibrosis may lead to ventricular dysfunction. More frequent cardiac monitoring is advised in the late, nonambulatory stage to reduce disease-related morbidity and mortality. Symptomatic heart failure can be particularly difficult to diagnose in nonambulatory patients with DMD. Clinical manifestations of heart failure: fatigue, sleep disturbance, and inability to tolerate daily activities are often unrecognized until late in the disease because of musculoskeletal limitations. The cardiologist should maximize medical therapy for heart failure. Consideration should also be given to thromboembolism prevention in patients with severe left ventricular dysfunction. Patients with DMD are at risk of rhythm abnormalities including atrial flutter or fibrillation, ventricular tachycardia, and ventricular fibrillation that can be treated with standard antiarrhythmic medications or device management. It is recommended to start annual Holter monitor screening with the onset of abnormal left ventricular function or development of myocardial fibrosis [3, 5].

Patients with DMD often have gastrointestinal or nutritional complications, including weight gain or loss, dietary or nutrient imbalance, fluid imbalance, swallowing dysfunction, and mandibular contracture. The purpose of nutritional care is to prevent overweight or obesity and undernutrition or malnutrition through regular assessment of growth and weight. It also helps to promote a healthy, balanced diet, with optimum intake of calories, protein, fluid, and micronutrients. It is recommended for the care team to include a registered nutritionist who should see an individual with DMD at every visit, beginning at diagnosis. More frequent monitoring by the nutritionist will be necessary during periods when weight gain or loss is anticipated. A physical therapist should be consulted to design exercise programs for patients who are at risk of becoming overweight. A speech-language pathologist should be consulted for patients with suspected dysphagia. A gastroenterologist needs to be consulted for problems with constipation, gastroesophageal reflux, and gastrointestinal motility concerns, and when gastrostomy tube placement is needed. Good nutritional status is defined as weight for length, or body-mass index (BMI) for age, that falls between the 10th and 85th percentiles on standard growth charts. If BMI cannot be calculated in patients with DMD, because height cannot be measured, weight-for-age percentiles should be used. Patients with DMD have altered body composition, so the use of standard growth charts is not optimal. Patients with DMD are at risk of overweight or obesity in childhood, with an increased risk of undernutrition or malnutrition as they approach adulthood. In early childhood, glucocorticoid therapy increases the risk of being overweight due to increased appetite and caloric intake and sodium and fluid retention. Loss of ambulation leads to decreased activity, which reduces caloric needs and puts patients at risk of becoming overweight. The clinician should create a nutritional plan that includes recommendations for calorie, protein, micronutrient, and fluid intake. If weight gain is excessive, an obesity management plan should be created, which addresses both diet and physical activity. Dysphagia is common and frequently progressive in patients with DMD. Screening questions should focus on perceived difficulty with swallowing, time necessary to eat an average meal, and interference of eating with quality of life. If a patient responds to screening questions in the affirmative, the speech-language pathologist should be consulted for an assessment, including a videofluoroscopic swallowing study. Patients often lose weight unintentionally before and during the onset of clinical symptoms of dysphagia. Their BMI or weight percentiles might decrease from the overweight category into the normal range or into the underweight range as a result of dysphagia and disease progression. Family and specialists should consider gastrostomy tube placement to be a necessary and positive intervention when progressive weakness interferes with self-feeding and swallowing. Malnutrition that is unresponsive to interventions to improve oral caloric intake, presence of moderate or severe dysphagia, and inability to maintain adequate hydration are indications for gastrostomy tube placement. Constipation is frequent complication of DMD. Risk factors include decreased colonic transit time, abdominal muscle weakness, and dehydration. Daily treatment with osmotic laxatives such as polyethylene glycol, milk, or lactulose might be necessary. Risk factors for gastro-esophageal reflux include esophageal dysmotility, delayed gastric emptying time, glucocorticoid therapy, and scoliosis. Treatment of gastroesophageal reflux consists of gastric acid suppression proton-pump inhibitors such as lansoprazole or omeprazole. Dietary access includes eating smaller and more frequent meals and decreasing dietary fat intake. As skeletal muscle weakness progresses in individuals with DMD, a delay in gastric emptying can occur, which can lead to postprandial abdominal pain, nausea, vomiting, and loss of appetite. Treatment options include dietary modification, pharmacological therapy, and postpyloric feeding with a gastrojejunal feeding tube [2].

The endocrine complications of DMD and its treatment include impaired growth, delayed puberty, and rarely, adrenal insufficiency. The goals of endocrine care are to monitor growth and development, identify and diagnose hormone deficiencies, and provide endocrine hormone replacement therapy when indicated. Impaired linear growth is common in patients with DMD and exacerbated by glucocorticoid treatment. Linear growth should be assessed every 6 months until completion of puberty. Standing height is the most appropriate measure in ambulatory patients. Height should be followed on a standardized growth curve. Growth monitoring with use of non-standing height measure should begin during the ambulatory stage to allow more accurate assessment after individuals lose ambulation. In nonambulatory patients, arm span, ulnar length, tibia length, knee height, and segmentally measured recumbent length have all been used to assess growth. A decline in growth trajectory, as evidenced by downward crossing of height percentile or an annualized height velocity of less than 4 cm per year, is consistent with impaired linear growth and indicates the need endocrinologist. Patients with a height of less than the third percentile should be referred. Assessment of impaired linear growth should include standard screening tests to assess for endocrine hormone or other abnormalities associated with growth failure. It is not recommended the routine use of recombinant human growth hormone to treat DMD-related growth failure. The decision to treat with recombinant human growth hormone should be based on a discussion of the potential risks and benefits of the therapy and reserved for patients with abnormal growth hormone stimulation test results. Delayed puberty due to hypogonadism is a potential complication of glucocorticoid therapy and can lead to be psychological distress. The absence of pubertal development by the age of 14 requires prompt referral to an endocrinologist. Biochemical testing using pediatric or ultrasensitive assays should be done to confirm the diagnosis of hypogonadism in patients with evidence of delayed puberty. A radiograph of the left hand for establishing bone age should also be considered. Testosterone replacement therapy is recommended in patients older than 14 years with confirmed hypogonadism. The potential benefits of testosterone on emotional and physical health usually outweigh the potential side effects, such as behavioral changes, acne, rapid growth spurt, and epiphyseal closure. Testosterone replacement, in order to mimic normal puberty, should be initiated at a low dose and slowly increased to adult replacement doses over several years. Intramuscular or topical preparations can be used. Testosterone concentrations should be monitored in all patients [2, 5].

6. Gene therapy

Gene therapy is the special treatment of genetic diseases through the delivery of corrective "therapeutic DNA" into the genetic material of a patient's cells. The corrective DNA is packaged within a vector that is used to effectively deliver the therapeutic DNA into targeted cells. The cell machinery uses the therapeutic DNA to produce functional proteins that the defective DNA could not, resulting in the correction of the underlying cause of disease pathology directly by restoring the lost function. Traditional drug-based approaches address disease symptoms but not the underlying genetic cause. Gene therapy directly targets the specific genetic defects that are the cause of the genetic disease. A major milestone in the treatment of DMD occurred in August 2014 with the registration of ataluren which was designed by PTC Therapeutics and registrated under the brand name Translarna™. Ataluren is an orally available, small molecule compound that targets nonsense mutations, and is the first drug in its class. It is indicated for the treatment of Duchenne muscular dystrophy resulting from a nonsense mutation in the dystrophin gene, in ambulatory patients aged 2 years and older. The presence of a nonsense mutation in the dystrophin gene should be determined by genetic testing [9]. Up to 15% of DMD cases are caused by this type of mutation. A nonsense mutation in DNA causes a premature stop codon within an mRNA. This premature stop codon in the mRNA causes disease by terminating translation before a full-length protein is generated. Ataluren enables ribosomal read through of mRNA containing such a premature stop codon, resulting in production of a full-length protein [10].

Before starting therapy with ataluren, the patient must be on glucocorticoid therapy for at least 6 months. Translarna was tested in a randomized, double-blind Phase 2b study (NCT00592553) conducted in 11 countries to assess the treatment's safety, dosage, and efficacy in males (5 years or older) with DMD caused by non-sense mutations. Results at 48 weeks showed that the treatment was generally well-tolerated and led to increases in the distance walked in six minutes, a standard test of functional capacity, compared to a placebo. Statistically significant benefits were found in timed function tests, such as the 10-meter run/walk test, the 4-stair climbing test, and the ability to stand when lying on the back or with the face upward. STRIDE (strategic targeting of registries and international database of excellence) is an ongoing observational global study (NCT02369731) — is following patients treated with Translarna, plus standard care (including corticosteroids), for at least five years. It is being conducted in partnership with the neuromuscular network TREAT-NMD and is currently recruiting participants [9].

The schedule of taking therapy is three times a day. The recommended dose is 10 mg (per kg of body weight) in the morning and at midday, and 20 mg/kg in the evening. The total recommended daily dose is 40 mg/kg. Recommended dosing intervals are six hours between morning and midday doses, six hours between midday and evening doses, and 12 h between the evening dose and the dose the following morning. Ataluren (Translarna™) has been given "conditional approval". This means that there is more evidence to come about this medicine [9, 10]. The European Medicines Agency will review new information on this medicine at least every year. Ataluren (Translarna™) has been approved in the European Union and Brazil. In 2020, the Committee for Medicinal Products for Human Use, part of the European Medicines Agency, recommended expanding the use of Translarna to DMD patients who lost their ability to walk.

In the U.S. however, the Food and Drug Administration rejected Translarna's approval, stating that it is unable to approve the therapy due to the lack of substantial evidence of its effectiveness [9, 11].

US Food and Drug Administration (FDA) has approved the use of eteplirsen, trade name Exondys 51®. Eteplirsen is a gene therapy, indicated for the treatment of persons with DMD as due to mutation in exon 51, which affects about 13% of individuals with DMD. It is used with patients whose mutations result in a shift in the reading frame and synthesis of dysfunctional dystrophins. Sarepta Therapeutics, a global US biotechnology company that specializes in genetics research and medications for rare diseases, is responsible for placing eteplirsen on the market. Eteplirsen is based on morpholine phosphorodiamidate oligomer (phosphorodiamidate morpholin oligomer PMO), an oligomer used to modify gene expression. PMOs are short, single-stranded analogs of the DNA molecule, constructed upon morpholine rings connected by phosphorodiamidate bonds, and bind to complementary target mRNA sequences. PMO allows skipping exon 51, and in that way, correct the reading frame. The result is a synthesis of shorter, but functional dystrophin. Eteplirsen is administered intravenously. On the same principle is based golodirsen, a medicine approved by the FDA in December 2019, a trade name Vyondys 53®, and the difference is that it is used in patients with a mutation in exon 53. Although there has been an increase in functional synthesis in several randomized clinical trials of dystrophin, the safety of eteplirsen is still being investigated. The European Medicines Agency has refused conditionally approve the placing of eteplirsen on the European market [12, 13].

Sarepta Therapeutics made three approved antisense oligonucleotide products that are used in patients with DMD: Exondys 51® (eteplirsen), Vyondys 53® (golodirsen), and Amondys 45® (casimersen). These medicines target patients with DMD who have a confirmed mutation to exon 51, exon 53, and exon 45. On 25 February 2021, casimersen received its first approval in the USA. The approval of casimersen is granted under the US FDA accelerated approval program. Now it is continuing in the phase 3 development for the treatment of DMD. The approval was based on increase in dystrophin production in skeletal muscle in patients with DMD treated with casimersen. Another clinical benefit must be examined and confirmed for continued approval [14].

Great hopes are put in recent studies, which have shown highly promising improvements in animal models with intravascular delivery of the engineered micro-dystrophin gene by adeno-associated virus (AAV). Several human trials are now started to advance AAV micro-dystrophin therapy to DMD patients. In April 2022, Sarepta Therapeutics presented interim findings from its Phase 2 clinical trial (Study 102) in DMD for its new gene transfer therapy, using the AAV method. SRP-9001 (delandistrogene moxeparvovec) is an adeno-associated virus (AAV) mediated gene therapy that delivers a micro-dystrophin-encoding gene to the muscles and is considered to be a curative treatment. The specific vector deployed in the gene transfer,

AAVrh74, has been shown to achieve the efficient delivery of micro-dystrophin to skeletal muscle with tolerable immunogenicity. Sarepta Therapeutics is responsible for global development and manufacturing of SRP-9001. In December 2019, Roche partnered with Sarepta to combine Roche's global reach, commercial presence, and regulatory expertise with Sarepta's gene therapy candidate for Duchenne to accelerate access to SRP-9001 for patients outside the United States. Sarepta owns exclusive rights to the micro-dystrophin gene therapy program initially developed at the Abigail Wexner Research Institute at Nationwide Children's Hospital [15].

Sarepta's Study SRP-9001-102 (Study 102) is a double-blind, 1:1 randomized, placebo-controlled clinical trial of SRP-9001 patients with Duchenne muscular dystrophy between the ages of 4 to 7. Study 102 uses SRP-9001 material and has two primary endpoints: micro-dystrophin expression at 12 weeks and change in NSAA total score at 48 weeks compared to placebo. Secondary endpoints include timed functional tests; micro-dystrophin expression measured by immuno-fluorescence fiber intensity; and micro-dystrophin expression measured by immuno-fluorescence percent dystrophin positive fibers. In Part 1, results from the treatment and placebo groups were compared 48 weeks following treatment. In Part 2, the study remained blinded to the participants and investigators, while all participants in the placebo group crossed over to active treatment and all participants were followed for another 48 weeks while safety and efficacy were evaluated. Participants will be evaluated for five years total after treatment. SRP-9001-treated participants from the placebo crossover group scored a statistically significant 2.0 points higher on the mean North Star Ambulatory Assessment at 48 weeks compared to propensity-score weighted external controls. Mean NSAA scores from these Part 2 participants improved 1.3 points from baseline for the SRP-9001 treated group and the NSAA scores in the external control group declined 0.7 points from baseline. Study is now undergoing and additional results will be shared at a future medical congress. CAP-1002 is an investigational cell therapy developed by Capricor Therapeutics. It is believed to be used in treatment of heart conditions, including cardiomyopathy, or disease of the heart muscle, linked to Duchenne muscular dystrophy (DMD). The lack of dystrophin in the heart muscle in patients with DMD causes cardiomyopathy, one of the leading causes of death in DMD patients. CAP-1002 consists of cardiosphere-derived cells (CDCs). They are progenitor cells capable of developing into mature heart cells. By releasing sacks of cellular material called exosomes, CDCs modulate immune cell activity to promote heart repair. The CDCs in CAP-1002 come from the heart tissue of a healthy donor. Preclinical studies with CAP-1002 in mouse models of DMD showed that CDCs were able to improve exercise capacity, heart function, and function of skeletal muscles. CAP-1002 also inhibited scarring, inflammation, and oxidative stress in the preclinical models. A Phase 1/2 clinical trial (NCT02485938) called HOPE (Heart Outcomes Prevention Evaluation)-Duchenne tested CAP-1002 in 25 male patients, ages 12 and up, with DMD-related cardiomyopathy. Most participants relied on a wheelchair and had substantial shoulder function impairment. All the participants in the open-label trial were given standard care, including corticosteroids, and 13 received one dose of CAP-1002 (75 million cells) administered directly into the heart [16].

Results from HOPE-Duchenne indicated that treatment with CAP-1002 reduced heart muscle scarring, and helped thicken the heart's left ventricle, which is crucial for pumping oxygenated blood through the body. Benefits were still evident after six months and at one year. Capricor launched a double-blind Phase 2 trial (NCT03406780) called HOPE-2 in 2018. The study enrolled 20 boys and young men with relatively advanced DMD, 80% of whom were unable to walk. Participants were randomly assigned to receive a placebo or CAP-1002 (150 million cells per infusion), given via infusion into the bloodstream every three months for a year. All were given standard steroid treatment. Results from HOPE-2 showed that, compared with a placebo, CAP-1002 significantly improved upper limb function, as assessed via a validated test called performance of the upper limb (PUL) 1.2. It also improved several measures of lung and heart health. For example, left ventricle ejection fraction, which is an assessment of how much blood the heart pushes out to the body with each pump, was significantly higher, by 4% on average, among CAP-1002-treated patients. Capricor is planning a potentially pivotal Phase 3 clinical trial called HOPE-3 (NCT05126758) to test the safety and effectiveness of CAP-1002 in DMD. The study, which is not yet recruiting, plans to l about 68 boys and young men with DMD, ages 10 and older, who have some difficulty walking. Participants will be given infusions of CAP-1002 (150 million cells per infusion) or a placebo every three months for a year. The study's main goal is to assess the effect of treatment on upper limb function [17].

During the period between May and July of 2020, the author and co-author conducted a research on topic ,clinical characteristics, diagnostic approach, and treatment of Duchenne muscular dystrophy in Dalmatia, which is also a part of master's thesis. The aim of the study was to assess timely recognition of clinical characteristics and implementation of a multidisciplinary approach according to the latest guidelines and to analyze clinical courses in patients on gene therapy. The research is a cross-sectional and retrospective study. Patients with dystrophinopathies (DMD and BMD) and their contacts were singled out by retrospective analysis of medical documentation. The patient's parents provided information on the patient's medical history and more recent medical documentation by phone call and electronic mail because it was a period of the coronavirus pandemic. Data on current age, age at diagnosis, types of mutations, methods of confirming the diagnosis, clinical picture, and therapy they are receiving have been extracted. Numerical parameters included: ejection fraction, spirometry findings, bone mineral density obtained by densitometry (all parameters were extracted by recent medical documentation), and body mass index (it was measured by patients' parents). Data on two patients with a missense mutation in the dystrophin gene are receiving ataluren gene therapy.

A total of 9 patients were included in the study, 8 with a diagnosis of Duchenne (DMD) and 1 with a diagnosis of Becker muscular dystrophy (BDM). The age at diagnosis of DMD ranged between 2 and 6.5 years, while in patients with BMD it was set at 11 years of age. In some patients, there is a delay in establishing an accurate diagnosis for several years (**Table 1**). According to the parents of the patients, a lot of time for establishing the diagnosis was lost due to the wrong focus on the liver diseases (elevated AST and ALT), as well as the attitudes that their children are just clumsy or lazy. Of the mutation types, 4 patients have deletions, 3 duplications, and 2 point mutations. Most diagnoses, in addition to MLPA analysis, included muscle biopsy. six patients are ambulatory, while 3 patients are dependent on wheelchair use. In all patients, there was normal ejection fraction, while the oldest subject in the study was the only one to have a pathological ultrasound of the heart with signs of cardiomyopathy. 7 patients have normal pulmonary function (it was measured by spirometry), while 2 developed chronic pulmonary insufficiency and used noninvasive ventilation methods. 6 patients are on glucocorticoid therapy and all of them are using deflazacort. An increased risk of pathological fractures due to decreased bone density is found in 3 patients due to densitometry results. 2 patients receive gene therapy with ataluren. 1 of them receives therapy for 8 months and is in ambulatory stage, while the other receive therapy for 6 years, no longer has the ability to move independently, but

Patient number	Diagnosis	Age at first symptoms (years)	Age at diagnosis (years)
1	DMD	1	2
2	DMD	5	5
3	DMD	3	3
4	DMD	3	6
5	DMD	3	6,5
6	BMD	10	11
7	DMD	1,5	3,5
8	DMD	5	5
9	DMD	4,5	4,5

DMD, Duchenne muscular dystrophy; BMD, Becker muscular dystrophy.

Table 1.
Presentation of patients with their diagnosis, age at first symptoms, and age at confirmed diagnosis.

his hand function is preserved. It is important to stress the preserved hand function for patient in study and generally in patients with DMD who are nonambulatory. Hand function is more complex than the leg function and allows patients greater independence that reflects both the physical and mental condition of patients. Patients with DMD, with basic progressive muscle weakness, also develop complications of other organ systems, especially respiratory and cardiac complications. The importance of multidisciplinary care in patients with DMD is manifested in the prevention of complications, prolongation of life expectancy, and raising the quality of life. Gene therapies are taking place in the treatment of DMD, and the results of studies of these causal therapies show encouraging results. The advent of gene therapy as a causal therapy for DMD has placed additional emphasis on diagnosing DMD as early as possible, due to earlier initiation of the treatment, an additional prolongation of life expectancy, and increased quality of life in patients with DMD [18].

7. Conclusion

Improvements in the function, longevity, and quality of life of patients with Duchenne muscular dystrophy (DMD) have been achieved through a multidisciplinary approach to management across a range of healthcare specialties.

Patients with Duchenne and Becker muscular dystrophy with basic progressive muscle weakness of the skeletal system, develop complications of numerous organ systems that significantly contribute to the deterioration of the clinical condition and shorter life expectancy. Timely diagnosis and multidisciplinary care prolonged the life expectancy of patients with DMD, and the development of subspecialist branches has enabled the improvement of diagnostic methods and treatments.

It is important to notice subtle signs of the disease (slowed motor development and speech development, etc.) in order to be as early as possible suspected DMD and confirmed the diagnosis as soon as possible.

It is certainly necessary to emphasize again the timely diagnosis, because every day the patient receives pharmacological and physical therapy contributes to longer life expectancy and better quality of life. Primary pediatric care or family medicine doctors are usually the first ones who meet patients with undiagnosed DMD. Thus, it is very important to emphasize their role in timely and accurate diagnosis of DMD and BMD. At any sign of muscle weakness, doctors should refer patients to laboratory test because the most important screening test for dystrophinopathies is determination of serum creatine kinase (CK). Gene therapy as a causal therapy for DMD is a major milestone in treatment and is considered a therapy that will become the mainstay in the treatment of dystrophies in the future. For now, gene therapy is available to treat certain types of mutations, such as nonsense mutations. Great hopes are placed in microdystrophin studies, which are currently underway.

In the last few years, therapeutic options in the treatment of DMD have advanced significantly, and new ones are emerging. The very fact that there is a causal therapy puts emphasis on early diagnosis and the earliest possible start with a therapy that provides much hope for success in a treatment of patients with DMD and BMD [5].

Author details

Radenka Kuzmanić Šamija[1]* and Marta Plejić[2]

1 University Hospital Center Rijeka, Croatia

2 Health Center of Split, Croatia

*Address all correspondence to: dadakuzmanicsamija@gmail.com

References

[1] Mardesic D. Neuromuscular diseases. In: Mardesic D, Baric I, editors. Pediatrics. Zagreb: Skolska knjiga; 2016. pp. 1016-1019

[2] Birnkrant DJ, Bushby K, Bann CM, Apkon SD, Blackwell A, Brumbaugh D. Diagnosis and management of Duchenne muscular dystrophy. Part 1: Diagnosis, and neuromuscular, rehabilitation, endocrine, and gastrointestinal and nutritional management. Lancet Neurology. 2018;**3**:251-267

[3] Birnkrant DJ, Bushby K, Bann CM, Alman BA, Apkon SD, Blackwell A. Diagnosis and management of Duchenne muscular dystrophy. Part 2. Respiratory, cardiac, bone health, and orthopaedic management. Lancet Neurology. 2018;**4**:347-361

[4] Gloss D, Moxley RT, Ashwal S, Oskoui M. Practice guideline update summary: Corticosteroid treatment of Duchenne muscular dystrophy: Report of the guideline development subcommittee of the American Academy of Neurology. Neurology. 2016;**5**:465-472. DOI: 10.1212/WNL.0000000000002337

[5] Birnkrant DJ, Buhsby K, Bann CM, Apkon SD, Blackwell A, Colvin MK. Diagnosis and management of Duchenne muscular dystrophy. Part 3: Primary care, emergency management, psychosocial care, and transitions of care across the lifespan. Lancet Neurology. 2018;**5**:445-455

[6] Bachrach LK. Osteoporosis and measurement of bone mass in children and adolescents. Endocrinology and Metabolism Clinics of North America. 2005;**34**:521-535. DOI: 10.1016/j.ecl.2005.04.001

[7] Bian Q, McAdam L, Grynpas M, Mitchell J, Harrington J. Increased rates of vitamin D insufficiency in boys with Duchenne muscular dystrophy despite higher vitamin D3 supplementation. Global Pediatric Health. 2019;**6**:1-7. DOI: 10.1177/2333794X19835661

[8] Gomez-Merino E, Bach JR. Duchenne muscular dystrophy: Prolongation of life by noninvasive ventilation and mechanically assisted coughing. American Journal of Physical Medicine & Rehabilitation. 2002;**81**:411-415. DOI: 10.1097/00002060-200206000-00003

[9] PtcBio.Com [Internet], South Plainfi eld, PTC Therapeutics Inc.: PTC Provides Update on Translarna™ (ataluren) Application for Label Expansion. 2020. Available from: https://ir.ptcbio.com/news-releases/news-release-details/ptc-provides-update-translarnatm-atalurenapplication-label

[10] Mercuri E, Muntoni F, Osorio AN, Tilinius M, Buccella F, Mongenroth LP. Safety and effectiveness of ataluren: Comparison of results from the STRIDE registry and CINRG DMD natural history study. Journal of Comparative Effectiveness Research. 2020;**5**:341-360. DOI: 10.2217/cer-2019-0171

[11] McDonald CM, Campbell C, Torricelli RE, Finkel RS, Flanigan KM, Goemans N. Ataluren in patients with nonsense mutation Duchenne muscular dystrophy (ACT DMD): A multicentre, randomised, double-blind, placebo-controlled, phase 3 trial. Lancet. 2017;**10101**:1489-1498

[12] Sarepta.Com [Internet], Cambridge, Sarepta Therapeutics: Gene Therapy Engine Building a Gene Therapy Engine for a Range of Rare Diseases. 2020. Available from: https://www.sarepta.com/science/genetherapy-engine

[13] National Library of Medicine [Internet]. Eteplirsen in the treatment of Duchenne muscular dystrophy. 2017. Available from: https://www.ncbi.nlm.nih.gov/pmc/articles/PMC5338848/

[14] Sarepta Therapeutics [Internet]. Sarepta Therapeutics Announces Fourth Quarter and Full-Year 2021 Financial Results and Recent Corporate Developments 2022. Available from: https://investorrelations.sarepta.com/news-releases/news-release-details/sarepta-therapeutics-announces-fourth-quarter-and-full-year-2021

[15] Duan. Systemic AAV micro-dystrophin gene therapy for Duchenne muscular dystrophy. Molecular Therapy. 2018. DOI: 10.1016/j.ymthe.2018.07.011

[16] Sarepta Therapeutics' Gene Therapy SRP-9001. External Control in Part 2 of Study SRP-9001-102 for the Treatment of Duchenne Muscular Dystrophy. 2022. Available from: https://investorrelations.sarepta.com/news-releases/news-release-details/sarepta-therapeutics-gene-therapy-srp-9001-shows-statistically

[17] The Lancet. Repeated Intravenous Cardiosphere-derived Cell Therapy in Late-stage Duchenne Muscular Dystrophy (HOPE-2): A Multicentre, Randomised, Double-blind, Placebo-controlled, Phase 2 Trial. 2022. Available from: https://www.thelancet.com/journals/lancet/article/PIIS0140-6736(22)00012-5/fulltext

[18] Kuzmanic SR, Plejic M. Clinical Characteristics, Diagnostic Approach and Treatment of Duchenne Muscular Dystrophy in Dalmatia. Croatia: University of Split; 2020

Chapter 5

Novel Insights into the Use of Biologicals in Idiopathic Inflammatory Myopathies

Ashwin Parchani, Aditya Sudan, Shiana Singh, Arpit Singh and Monika Pathania

Abstract

Idiopathic inflammatory myopathies (IIMs) are a set of autoimmune disorders characterized by muscle inflammation and weakness, as well as a variety of extra-muscular presentations. IIMs are remarkably complex and difficult to treat, and glucocorticoid treatment and synthetic immunosuppressants are frequently ineffective. The pathophysiology of IIM has been linked to defects in both the innate and adaptive immune systems. Multiple prospective targets for biologic therapy have been studied because of a greater understanding of the main cytokines, as well as the cell-mediated and antibody effectors of disease. B-cell depletion with rituximab, as well as tumor necrosis factor inhibitors and other biologic treatments, is among the most extensively studied drug in IIM. There is currently no straightforward way to define all of the pharmaceuticals that are classified as biologics. This group of drugs has gained a lot of interest in the recent era for the treatment of various autoimmune and skeletal muscle disorders. This chapter shall address the mechanism of action, side effects, uses, and scope of biologics used in treatment of IIM.

Keywords: dermatomyositis, polymyositis, idiopathic inflammatory myopathy, biologics, rituximab

1. Introduction

The idiopathic inflammatory myopathies (IIMs)/myositis syndromes are a heterogeneous group of systemic autoimmune conditions that include polymyositis, dermatomyositis (DM), necrotizing myopathy, inclusion body myositis (IBM), anti-synthetase syndrome, and overlap syndromes with myositis. These have a significant influence on skeletal muscle, though they can also have extra-muscular consequences. They are linked to considerable disability as a result of progressive weakness, as well as an increased risk of mortality. These clinical signs, along with muscle biopsy data and specific serum autoantibodies, are used to make the diagnosis.

IIMs have always been difficult to treat. Glucocorticoids and traditional immuno-suppressive or immunomodulatory drugs such as methotrexate, azathioprine, myco-phenolate mofetil, tacrolimus, cyclosporine, and intravenous immunoglobulin are examples of traditional treatment modalities. Some patients have recurrences of the disease during or after conventional therapy, while others do not respond completely, which can pose therapeutic challenges. A considerable number of patients have a partial response, necessitating long-term glucocorticoid therapy, which has its own set of adverse effects as well as the implications of incomplete disease control, such as persistent muscle deterioration. As a result, there has been an increasing interest in evaluating innovative and targeted therapeutics, such as biologics, that target specific pathways involved in IIM etiopathogenesis.

Biomarkers linked to IIM pathogenesis have been investigated utilizing a range of approaches, including cytokine/chemokine investigations, enhanced immunohistochemistry and flow cytometry, microarrays, and RNA-sequencing analysis. Multiple potential targets for biologic therapy have been identified because of growing knowledge of important cytokines as well as cell-mediated and antibody effectors of disease.

The introduction of biologic therapies has held promising potential for autoimmune diseases, allowing us to translate our knowledge of specific disease pathophysiology processes into medications that target certain autoimmune disorders. The aim of this chapter is to outline the pharmacologic profile of biologic treatment of myositis as per currently available literature.

2. Mechanism of action

2.1 Biological DMARDS – mechanisms of action

The complex pathogenesis of rheumatic and musculoskeletal disorders has gradually been pieced together, and this has led to the appreciation of the underlying cytokine networks underlying these disorders [1]. This has resulted in the development of targeted biological therapies with myriad mechanisms of action. For the purposes of understanding this broad and heterogeneous topic, the therapeutic agents will be classified, albeit arbitrarily based on the primary biological signaling pathways being targeted.

2.2 Biological targeting TNF-alpha signaling

Tumor necrosis factor-Alpha (TNF-Alpha) is produced by a wide variety of both immune and non-immune cells. It exists as a 26 kD transmembrane protein (tmTNF-Alpha), which is cleaved by the extracellular metalloproteinase, ADAM-17/TACE, which results in the release of a soluble form of TNF-Alpha (sTNF-Alpha). Both the transmembrane and soluble versions of TNF-Alpha are biologically active and signal via the two distinct TNF-Alpha receptors – TNFR1 and TNFR2 [2]. The receptors have partially redundant but distinct downstream signaling cascades, which result in differences in biological function, which have been highlighted in **Figure 1** [3]. TNFR1 signals via the canonical NF-κB pathway and may be pro-inflammatory, pro-survival, or pro-apoptotic in the given immunological context. TNFR2 signals via both the canonical and non-canonical NF-κB signaling cascades but lacks the pro-apoptotic signaling demonstrated by TNFR1. The biological outcomes of signaling via these two receptors are best exemplified by the effects they have on T-Regulatory (Treg) cell survival—TNFR1 enhances Treg cellular apoptosis while TNFR2 (as it lacks a death domain unlike TNFR1) enhances the expression of the Treg cell master transcription

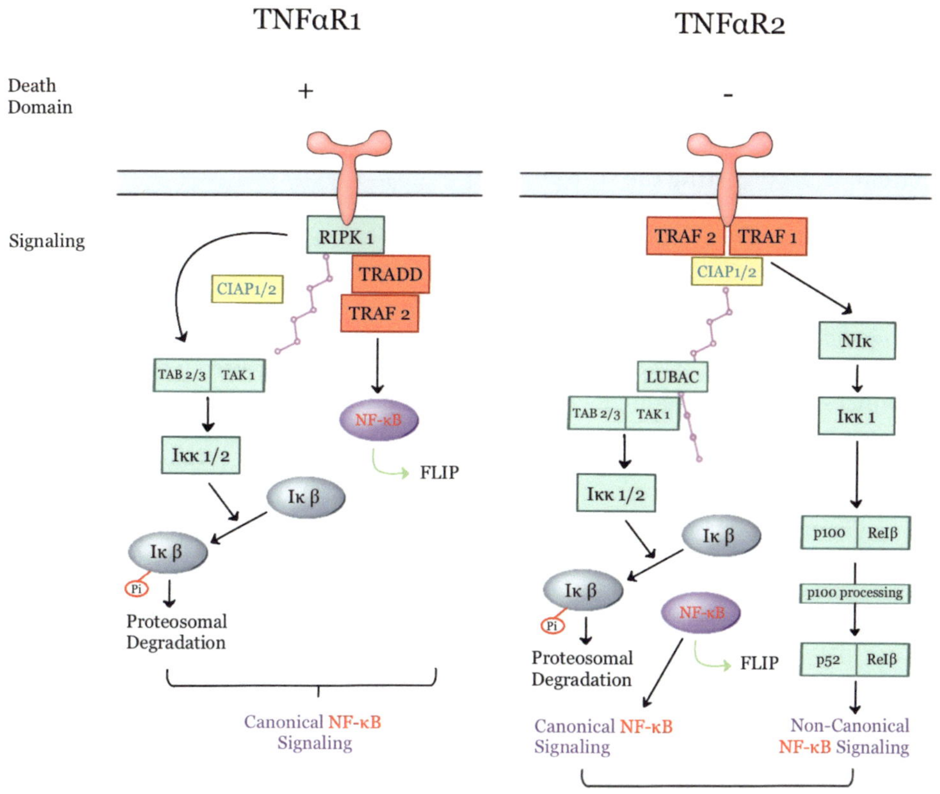

Figure 1.
Differences in biological functions of TNFR1 and TNFR2.

factor FOXP3, thus maintaining these regulatory cell populations. An additional difference between the two receptors of TNF-Alpha is that TNFR2 is able to interact with the tmTNF-Alpha, resulting in bidirectional signaling (both forward and backward), thus potentiating the immunoregulatory functions of TNFR2 signaling [4].

The therapeutic effects of anti-TNF-Alpha, although slightly variable based on the exact agent used, generally capitalize on the central role TNF-Alpha plays in determining pro vs. anti-inflammatory signaling. Overall biologicals targeting TNF-Alpha likely produce a clinical response as a result of the following effects:

1. Bind to sTNF-Alpha and counteract pro-inflammatory signaling via TNFR1.

2. Enhance apoptosis of pro-inflammatory cells- Possibly by blocking tmTNF-Alpha interactions with TNFR2 and/or enhancing the pro-apoptotic signaling downstream of TNFR1.

3. Direct antibody-dependent cellular cytotoxicity (ADCC) and complement-mediated cellular damage in cells expressing TNF-alpha [5].

A distinction must be made between etanercept and other anti-TNF-alpha agents in that is a fusion protein composed of the soluble portion of the TNFR2 and the constant region of the IgG1 molecule, while the latter are all monoclonal antibodies directed against TNF-alpha. Thus etanercept acts as a decoy receptor and is effective in cases of TNF-alpha receptor associated auto-inflammatory syndrome patients with the T50M mutation (who have been reported to not respond to infliximab therapy in some cases) [6].

2.3 Biologicals targeting the IL-6 pathway

Interleukin-6 (IL-6) is a member of the IL-6 superfamily of interleukins, along with leukemia inhibitory factor and oncostatin M—all three signal via receptors that contain the common gp130 subunit. IL-6 (like IL-11, IL-13, and IL-27) signals using JAK1/2 and TYK-2, which then phosphorylate and cause the nuclear translocation of STAT3 and STAT6 [7]. IL-6 signaling plays a central role in activating the systemic inflammatory response inducing acute phase reactant production in the liver, inducing megakaryocytic differentiation, and resetting the hypothalamic set point to cause fever [8]. Additionally, IL-6 signaling plays a critical role in determining TH17 vs. Treg cell polarization—IL-6 signaling suppresses FoxP3 expression (decreasing Treg polarization) and in the presence of concomitant TGF-Beta signaling inducing ROR-GammaT expression (enhancing T17 polarization) [9]. Blocking IL-6 signaling using targeted biologicals is therefore therapeutically very useful given the critical contribution of excessive innate immune activation and TH17 polarized T cells to the pathogenesis of RMDs such as rheumatoid arthritis. This is achieved clinically by targeting either IL-6 itself (Siltuximab, Clazakizumab) or the IL-6 Receptor (Tocilizumab, Sarilumab).

2.4 Biologicals targeting the type 1 interferon pathway

An appreciation of the central role played by type 1 interferons in the pathogenesis of diseases such as SLE and type 1 interferonopathies has resulted in the development of biologicals targeting IFN signaling [10, 11]. Anifrolumab is an anti-interferon alpha R1 (IFNAR1) monoclonal antibody, which has shown promise in the management of SLE. It downregulates the expression of IFNAR1 on various cell types with a resultant decrease in the phosphorylation and nuclear translocation of STAT1 (**Figure 2**).

2.5 Biologicals targeting the IL-17/IL-23 axis

As detailed above in the section on biologicals targeting IL-6 signaling, IL-23 and IL-6 play central roles in determining TH17 vs. Treg T cell polarization. TH17 cells as the name suggests in turn produce IL-17 (along with other innate lymphocytes that also express the master transcription factor ROR-γT) [12, 13]. Excessive TH17 polarization has been shown to be crucial to the pathogenesis of seronegative spondyloarthropathies and psoriasis. Molecules that target IL-23 signaling may target the p40 subunit of the IL-23R such as ustekinumab and briakinumab, or they target the p19 subunit of the same receptor such as guselkinumab, rizankizumab, and tidrakizumab. Biologicals targeting IL-17 usually target IL-17A/F such as secukinumab and ixekinumab [14]. Brodalumab is, however, different in that it targets the IL17A receptor [15].

2.6 Biologicals targeting immune checkpoint signaling

Abatacept is a fusion protein consisting of the extracellular domain of CTLA-4 (cytotoxic T-lymphocyte-associated protein 4) and the constant region of IgG1.

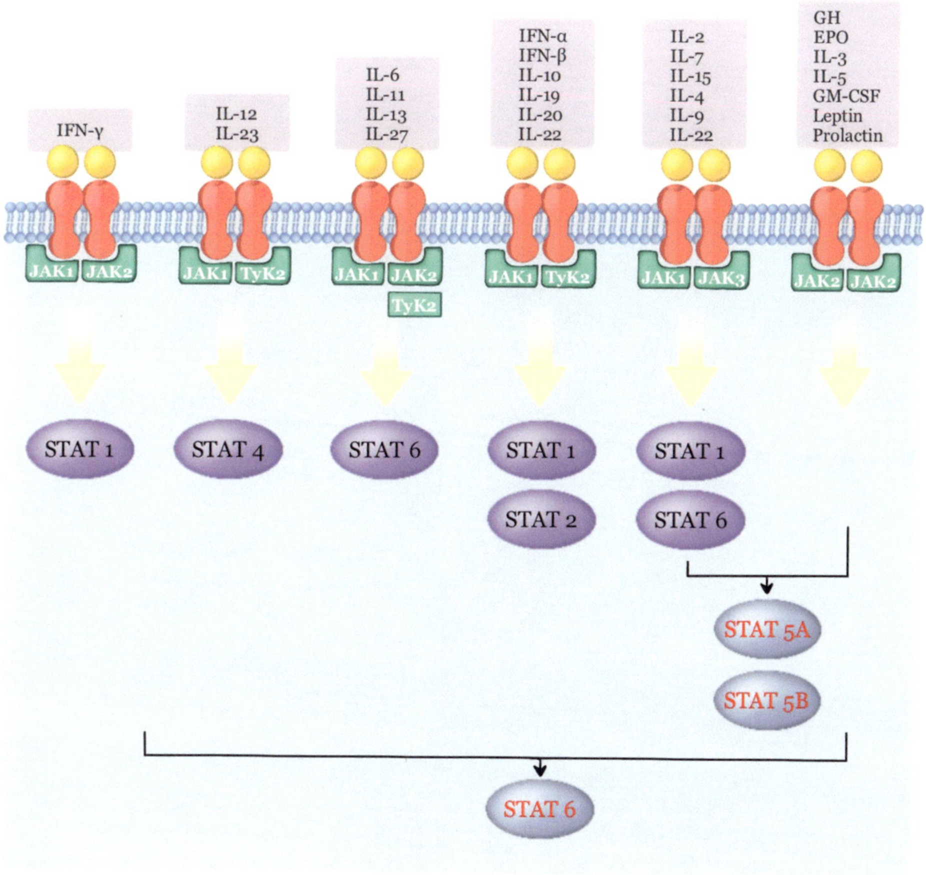

Figure 2.
Signaling pathways for various interferons. JAK – Janus kinase, TYK – tyrosine kinase, STAT – signal transducer and activator of transcription.

It prevents a co-stimulatory signal by binding to CD80/86 and preventing its interactions with CD-28 from being delivered to T- cells during antigen presentation, thus preventing activation of naïve T-cells [16]. In the context of autoimmune disorders, this likely prevents aberrant activation of partially/completely self-reactive T cells. This strategy has been of particular use in the management of rheumatoid arthritis (**Figure 3**) [17].

2.7 Biologicals targeting specific cell types

The phenotypic heterogeneity of various immune cell types allows for the highly specific targeting of various cells that play crucial roles in immune responses, using molecules developed against specific cell surface targets. Examples of this approach include Anti-CD20 specific biologicals, such as rituximab, which are able to deplete B cell numbers reliably and thus are effective if diseases where B cells play a central role- B-cells act as important antigen-presenting cells in the RA joint, IgG4RD, and as sources of autoantibodies in AAVs. Similarly, recently the SLAMF7 and CD38 targeting elotuzumab and darutumumab, respectively, have shown promise in the management of SLE [18].

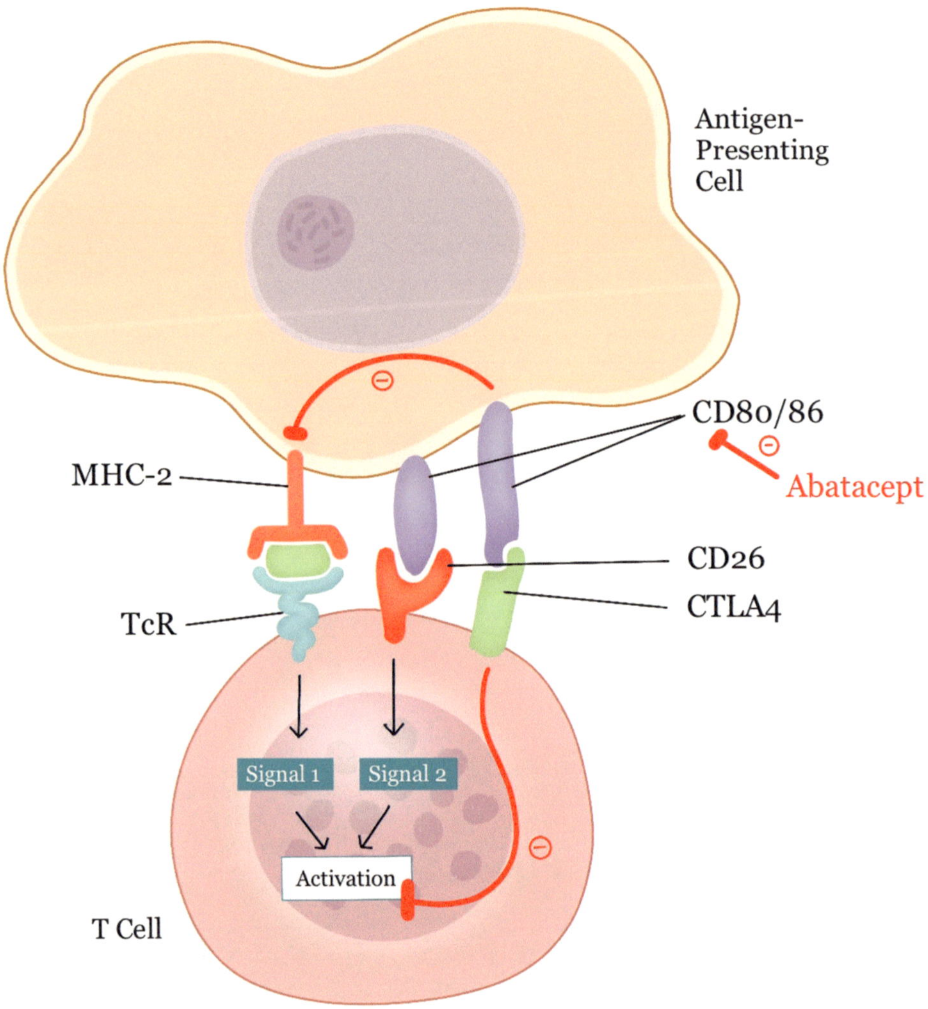

Figure 3.
Mechanism of action of abatacept. MHC-2 – major histocompatibility complex – 2, TcR – T cell receptor.

3. Pharmacology of biologics

The pharmacokinetics and pharmacodynamics of biological therapies, including monoclonal antibodies, are unique. Monoclonal antibodies have a distinct advantage over other drugs in that they can precisely bind and target certain antigen ligands with high affinity. These biologics, however, have some limitations in terms of pharmacological or pharmacokinetic features. They are absorbed, transported, and removed through completely distinct mechanisms, which poses significant challenges to how these drugs are delivered and can reach their pharmacological target [19–21].

Absorption: Because of the high molecular size of biologics and their breakdown in the gastrointestinal tract, most biologics are taken by the parenteral route, which includes intravenous (IV), subcutaneous (SC), and intramuscular (IM) injection [19, 20]. Bioavailability after SC and IM treatment might be variable, ranging from 20% to 95%.

Absorption following these routes of administration can be extremely slow (peak plasma concentrations reported 1–8 days after the dosage) and occurs mostly through the lymphatic system.

Distribution: In terms of distribution and tissue infiltration, biologicals can readily travel from the SC space by diffusion and/or convection through lymphatic capillaries. Pinocytosis or receptor-mediated endocytosis can allow them to reach intracellular destinations beyond systemic circulation [22]. The distribution of monoclonal antibodies (mAbs) in the vascular and interstitial fluids is explained by their large size and physicochemical features (charge and hydrophobicity). Tissue distribution accounts for 5–15% of the overall quantity of mAb, and distribution into the brain is quite limited (0.1%) [23]. If mAb-tissue target binding occurs with high affinity, a large proportion of mAb may be detected in the body.

As a result, mAbs might have high apparent volumes of distribution in steady state (Vss) [20].

Metabolism and Elimination: In mAb disposal, two metabolic routes, specific and nonspecific, are implicated, and their influence varies over time depending on the amount of free mAb in the plasma and the dosage provided. Metabolism through the reticuloendothelial system by pinocytosis/proteolysis reflects the linear and nonspecific clearance, which may be significant at certain dosage levels because of the higher endothelial surface area in the stomach, muscle, and skin [24]. The specific pathway begins once the receptor–drug combination is internalized, allowing the drug to enter the cell and be inactivated by cytoplasmic endosomes. FcRn, on the other hand, may bind IgG and mAbs at the acidic pH of the lysosome, avoid proteolysis, and return to the cell membrane [25–27].

3.1 Anti-TNF α

3.1.1 Etanercept

- Pharmacodynamics

Etanercept is a fully humanized, dimeric fusion protein made up of two copies of the extracellular ligand-binding region of the human TNF p75 receptor coupled to a part of immunoglobulin G1. It binds to TNF, preventing it from binding to cell surface receptors and inhibiting its pro-inflammatory effects [28].

- Pharmacokinetics

Absorption: Population pharmacokinetic modeling in adults with RA, AS, or who were healthy showed a subcutaneous bioavailability of 56.9% with a Ka of 0.0223/h [28].

Distribution: In adults with RA, population pharmacokinetic modeling predicts a total Vd of 5.49 L with a peripheral compartment of 1.24 L and an apparent Vd of 7.88 L after subcutaneous dosing in pediatric patients with JIA [28, 29].

Metabolism and Excretion: As etanercept is a fusion protein antibody, it is assumed to be metabolized and degraded via proteinases similarly to endogenous proteins.

Half-Life: 102 hours [30].

Clearance: 160 mL/h [30].

- Adverse Effects:
  - Infection (including bacterial infection, fungal infection, serious infection, viral infection: 50–81%)
  - Respiratory tract infection (21–54%), upper respiratory tract infection (38–65%)
  - Injection site reaction (adults: 15–43%; children: 7%; mild to moderate; usually decreases with subsequent injections)
  - Antibody development (non-neutralizing; 4–16%).
  - Diarrhea (3–16%).
  - Skin rash (3–13%).

3.1.2 Infliximab

- Pharmacodynamics

Infliximab inhibits the activation of the pro-inflammatory signaling cascade. Infliximab has been reported to prevent inflammatory cell infiltration into inflammatory areas. It also suppresses the expression of molecules involved in cellular adhesion, such as E-selectin, intercellular adhesion molecule-1 (ICAM-1) and vascular cell adhesion molecule-1 (VCAM-1), chemoattraction (IL-8 and monocyte chemotactic protein (MCP-1)), and tissue degradation (matrix metalloproteinase (MMP) 1 and 3).

- Pharmacokinetics

Absorption: Infliximab absorption follows a linear relationship between the dose given and the maximal serum concentration after a single intravenous infusion.

Distribution: The distribution at steady state was independent of dose in adult patients' pharmacokinetic investigation, indicating that infliximab was distributed largely within the vascular compartment.

Half-life: 7–12 days [31].

Clearance:11–15 mL/hour [31].

- Adverse Effect:
  - Infection (27–74%), serious infection (3–60%)
  - Antibody development (10–52%), increased ANA titer (~50%)
  - Abdominal pain (12–26%)
  - Nausea (21%)
  - Infusion-related reaction (≤20%)

- Headache (18%)
- Abscess (≤15%)
- Anemia (≤11%)

3.1.3 Adalimumab

- Pharmacodynamics

After treatment with adalimumab, a decrease in levels of acute-phase reactant proteins of inflammation (C reactive protein [CRP] and erythrocyte sedimentation rate [ESR]) and serum cytokines (IL-6) was measured compared with baseline in patients diagnosed with rheumatoid arthritis. CRP levels were also shown to be lower in Crohn's disease patients. After treatment with adalimumab, serum levels of matrix metalloproteinases (MMP-1 and MMP-3) that cause tissue remodeling and cartilage degradation were found to be lower [32]. In adult and pediatric patients with diverse inflammatory disorders, a reduction in disease signs and symptoms, induction of clinical response, suppression of structural damage, and improvements in physical function have been documented [33, 34].

- Pharmacokinetics

Absorption: Following a single 40 mg subcutaneous injection of adalimumab to healthy adult volunteers, the maximum serum concentration (Cmax) and time to achieve the maximum concentration (Tmax) were 4.7 ± 1.6 g/mL and 131 ± 56 hours, respectively. The average absolute bioavailability of adalimumab after a single 40 mg subcutaneous dose was 64%, according to three clinical investigations.

Distribution: The distribution volume (Vss) ranges from 4.7 to 6.0 L.

Half-life: The average terminal half-life was 2 weeks, ranging from 10 to 20 days in different trials.

Clearance: 12 mL/hr. [RA patients with dose 0.25–10 mg/kg].

- Adverse Effect:
    - Injection site reaction (5–20%)
    - Antibody development (3–26%)
    - Upper respiratory tract infection (17%)
    - Increased creatine phosphokinase in blood specimen (children and adolescents: 15%)
    - Positive ANA titer (12%)
    - Skin rash (12%)
    - Headache (12%)
    - Sinusitis (11%)

3.2 IL-1 Inhibitors

3.2.1 Anakinra

- Pharmacodynamics

Anakinra is a recombinant human interleukin-1 receptor antagonist (IL-1Ra) that inhibits capacity of interleukin-1 (IL-1) to bind to the IL-1 type I receptor (IL-1RI), therefore blocking its biologic function [35].

- Pharmacokinetics

Absorption: The bioavailability of anakinra is 95% in healthy subjects administered a 70 mg subcutaneous bolus injection.

Distribution: In adult subjects with rheumatoid arthritis (RA) treated with anakinra (n = 35), the volume of distribution averaged 18.5 L [36].

Elimination: Elimination is largely through the kidney, thus persons with compromised renal function are at risk to toxicity.

Half-life: In patients with rheumatoid arthritis (RA), the terminal half-life of anakinra ranged from 4 to 6 hours.

Clearance: Clearance is varied and rises with increasing creatinine clearance and body weight rise. The mean plasma clearance of anakinra was 16% and 50% lower in individuals with mild (creatinine clearance 50–80 mL/min) and moderate (creatinine clearance 30–49 mL/min) renal impairment, respectively. The mean plasma clearance of anakinra was 70% and 75% lower in patients with severe renal insufficiency and end-stage renal disease, respectively.

- Adverse Effect:
    - Injection site reaction (adults: 71%; infants, children, and adolescents: 16%)
    - Antibody development (49%)
    - Infection (39%)
    - Vomiting (14%)
    - Headache (12–14%)
    - Arthralgia (12%)

3.3 IL-6 Inhibitors

3.3.1 Tocilizumab

- Pharmacodynamics

Tocilizumab binds soluble and membrane-bound IL-6 receptors, preventing IL-6-mediated inflammation [37].

- Pharmacokinetics

Absorption: A 162 mg subcutaneous dose given weekly has a Cmax of 51.3 ± 23.2 µg/mL and an AUC of 8254 ± 3833 µg*h/mL [38].

Distribution: Tocilizumab is eliminated from the circulation in two phases after intravenous administration. The core volume of distribution was 3.5 L, and the peripheral volume of distribution was 2.9 L in rheumatoid arthritis patients, resulting in a volume of distribution of 6.4 L in steady state.

Half-life: Tocilizumab has a concentration-dependent half-life. In rheumatoid arthritis sufferers, the terminal half-life is 21.5 days.

Clearance: Clearance is dose-dependent, changing from nonlinear to linear at higher doses.

- Adverse Effect:

    - Injection site reaction (SubQ: children and adolescents: 15–44%; adults: 7–10%)
    - Increased serum alanine aminotransferase (≤36%), serum aspartate aminotransferase (≤22%)
    - Neutropenia (26–4%)
    - Increased serum cholesterol (19–20%)
    - Infusion-related reaction (4–20%)
    - Constipation (6–13%)
    - Arthralgia (12%)

3.4 IL-17 Inhibitors

3.4.1 Secukinumab

- Pharmacokinetics

Dosing: Secukinumab is administered by monthly subcutaneous injection after several loading doses [39].

Onset of action: Psoriasis: After 12 weeks, the optimal response can be established. (AAD-NPF [Menter 2019]).

Distribution: Vd: 7.1–8.6 L.

With increasing body weight, clearance and volume of distribution also increase.

Metabolism: Expected to be degraded into small peptides and amino acids via catabolic pathways similar to that which is seen with endogenous IgG.

Bioavailability: 55–77%.

Half-life elimination: 22–31 days.

Time to peak: ~6 days.

- Adverse Effect:
  - Infection (29–48%, serious infection, ≤1%).
  - Nasopharyngitis (11–12%).
  - Urticaria.
  - Hypercholesterolemia.
  - Diarrhea.

3.4.2 Ixekizumab

- Pharmacokinetics

Onset of action: Psoriasis: After 12 weeks, the optimal response can be established. (AAD-NPF [Menter 2019]).
Distribution: Vdss: 7.1 L. With increasing body weight, clearance and volume of distribution also increase.
Metabolism: Broken into tiny peptides and amino acids by catabolic processes similar to endogenous IgG.
Bioavailability: 60–81%.
Half-life elimination: 13 days.
Time to peak: ~4 days.

- Adverse Effect:
  - Neutropenia (11%, grades ≥3, <1%).
  - Antibody development (5–22%, neutralizing antibodies associated with decreased drug concentration and loss of efficacy, 2%).
  - URTI.
  - Conjunctivitis.

3.5 IL-12/23 Inhibitors

3.5.1 Ustekinumab

- Pharmacokinetics

Onset of action: Psoriasis: After 12 weeks, the optimal response can be established. (AAD-NPF [Menter 2019]).
Half-life elimination: SubQ: 14.9 ± 4.6 to 45.6 ± 80.2 days.
Time to peak: Psoriasis: SubQ: 45 mg: 13.5 days; 90 mg: 7 days.

- Adverse Effect:
  - Antibody development,

- Infections
- Nasopharyngitis

3.5.2 Guselkumab

- Pharmacokinetics

Onset of action: Psoriasis: After 12 weeks, the optimal response can be established. (AAD-NPF [Menter 2019]).
Distribution: Vd: 13.5 L.
Metabolism: Degraded into small peptides and amino acids via catabolic pathways in the same manner as endogenous IgG.
Bioavailability: Subcutaneous: ~49%.
Half-life elimination: 15–18 days.
Time to peak: 5.5 days.

- Adverse Effect:
  - Antibody development
  - Infections
  - Nasopharyngitis

3.6 Costimulation blockade

3.6.1 Abatacept

- Pharmacodynamics

CTLA-4 with the Fc component of immunoglobulin G1 (IgG1) forms Abatacept, a soluble fusion protein (CTLA4-Ig). It can be used in the treatment of rheumatoid arthritis, juvenile idiopathic arthritis, and psoriatic arthritis. Following numerous loading doses, abatacept can be given either as a weekly subcutaneous injection or as a monthly intravenous infusion.

- Pharmacokinetics

Distribution: IV: 0.07 L/kg (range: 0.02–0.13 L/kg).
Bioavailability: Subcutaneous: 78.6% (relative to IV administration).
Half-life: IV: 13.1 days (range: 8–25 days).
Clearance: Increases with increasing body weight.

3.7 Anti B-cell depletion and inhibition

3.7.1 Rituximab

- Pharmacokinetics

Onset of action: Within 2 weeks.
Duration: Depletion of B cells lasts at least 6 months.
Distribution: 3.1 L.
Half-life elimination: 18 days (range:5–78 days).
Clearance: 0.335 L/day.

- Adverse Effect:
  - Fatal infusion-related reactions.
  - Mucocutaneous reactions.
  - Hepatitis B virus (HBV) reactivation.
  - Progressive multifocal leukoencephalopathy (PML) (**Table 1**).

Drug	Bio-availability	Volume of Distribution (L/kg)	Half-life (units shown)	Clearance (mL/hour)	Adverse Effects
TNF-α Inhibitors					
Etarnecept	56.9%	5.49–7.88	102 hours	160	Infection, Injection site reaction, Antibody development, Diarrhea, Skin Rash
Infliximab	N.R.	N.R.	7–12 days	11–15	Infection, Antibody development, Abdominal pain, Nausea, Infusion-related reaction, Headache, Abscess, Anemia
Adalimumab	64%	4.7–6.0	10–20 days	12	Injection site reaction, Antibody development, Upper respiratory tract infection, Increased creatine phosphokinase in a blood specimen, Positive ANA titer, Skin rash, Headache, Sinusitis
IL-1 Inhibitors					
Anakinra	95%	18.5	4–6 hours	N.R.	Injection site reaction, Antibody development, Infection, Vomiting, Headache, Arthralgia
IL-6 Inhibitors					
Toculizumab	N.R.	6.4	21.5 days	N.R.	Injection site reaction, Increased S. ALT, S. AST, Neutropenia, Increased S. Cholesterol, Infusion-related reaction, Constipation

Drug	Bio-availability	Volume of Distribution (L/kg)	Half-life (units shown)	Clearance (mL/hour)	Adverse Effects
IL-17 Inhibitors					
Secukinumab	55%–77%	7.1–8.6	22–31 days	N.R.	Infection, Nasopharyngitis, Urticaria, Hypercholesterolemia Diarrhea
Ixekizumab	60%–81%	7.1	13 days	16.5	Neutropenia, Antibody development, URTI, Conjunctivitis
IL-12/23 Inhibitors					
Ustekinumab	57.2%	0.076–0.161	14.9 to 45.6 days	7.91	Antibody development, Infections, Nasopharyngitis
Guselkumab	49%	13.5	15 to 18 days	21.5.	Antibody development, Infections, Nasopharyngitis
Costimulation blockade					
Abatacept	78.6%	0.07	8 to 25 days	Adults: 0.22 mL/hr/kg Children: 0.4 mL/hr/kg	Hypertension Nausea Anemia Antibody development
Anti B-cell depletion and inhibition					
Rituximab	N.R.	3.1	5 to 78 days	0.335 (L/day)	Fatal infusion-related reactions, Mucocutaneous reactions, hepatitis B virus (HBV) reactivation, Progressive multifocal leukoencephalopathy (PML)

Table 1.
Pharmacokinetics parameters of biologicals.

4. Clinical uses

4.1 Rituximab

Rituximab is a chimeric monoclonal antibody that targets the CD20 antigen on the surface of B cells, causing them to be depleted in the bloodstream. Several case reports, case series, and open-label trials have demonstrated that rituximab can benefit refractory myositis patients [40]. The rituximab in myositis (RIM) trial, which included 195 subjects who were refractory to glucocorticoids and at least one immunosuppressive agent, is the largest randomized, double-blind, placebo-controlled trial on the efficacy of rituximab in adult and juvenile myositis to date [41]. Despite the fact that the primary goal was not met, the majority of patients (83%) experienced clinical improvement and steroid-sparing during the trial. The rituximab treatment was generally well tolerated, with infections being the most common side effect. Anti-Jo1 and anti-Mi-2 antibodies were found to be predictive of a successful response to rituximab in a post-hoc analysis of the RIM trial [42]. After B-cell depletion, both antibody levels dropped, and this was linked to changes in disease activity [43]. In a registry-based research of 43 individuals with ASS, the efficacy of rituximab was further assessed by comparing the clinical response after numerous rituximab cycles in antisynthetase antibody-positive and -negative patients [44]. Only the antibody-positive group demonstrated a significant steroid-sparing impact, though both groups showed clinical improvement regardless of antibody status.

A recent retrospective cohort analysis of 43 patients with refractory myositis found that rituximab is effective, with 75% of patients showing clinical and laboratory improvement after 1 year, as well as a considerable reduction/discontinuation of glucocorticoids [45].

Although the role of B lymphocytes in the development of myositis is not fully known, the present literature supports the use of rituximab in patients with refractory myositis. Infusion responses, potential cardiotoxic effects, and serious infections are all common side effects of rituximab treatment.

4.2 Anakinra

Anakinra, an IL-1 receptor antagonist, was explored in 15 individuals with refractory myositis in a small case study [46]. Seven patients responded clinically to the core set of disease activity measures used by the International Myositis Assessment and Clinical Studies (IMACS), and four of them improved their functional index scores. Randomized controlled research is still required to corroborate the findings.

4.3 Anti-TNFα therapies

Monoclonal antibodies, such as infliximab, and circulating receptor fusion proteins, such as etanercept, are examples of TNFα inhibitors. The present evidence in the literature for anti-TNFα therapy in myositis is mixed, with some studies and trials indicating a positive benefit in myositis patients, while others indicate no efficacy or even worsening symptoms following TNFα inhibitor treatment [47]. Only four out of 12 patients responded to therapy with infliximab in a recent randomized, double-blind, placebo-controlled trial investigating the efficacy of infliximab in refractory PM and DM [48]. However, two individuals who looked to respond to infliximab had their myositis worsen, and restarting the drug was linked to anaphylaxis and the formation of anti-dsDNA antibodies [49]. Infliximab usage was related to better muscle strength and fatigue but only a partial decline in blood CK levels in a larger retrospective study of eight patients with refractory dermatomyositis or polymyositis [50]. Infliximab therapy (four infusions of 5 mg/kg body weight over 14 weeks) proved ineffective in a more recent pilot study of 13 individuals with refractory IIM [51]. A multicenter, open-label, controlled trial of infliximab in combination with weekly methotrexate in patients with polymyositis or dermatomyositis was prematurely terminated due to a low inclusion rate and a high drop-out rate due to disease progression and infusion reactions [52]. TNFα inhibitors may also cause myositis, according to some reports [53]. Therefore, the use of TNFα inhibitors in the treatment of myositis cannot be supported at present.

4.4 Tocilizumab

Tocilizumab, an interleukin 6 (IL-6) receptor antagonist, has only been used in a few case reports so far; the first involved two patients with refractory Jo-1-positive PM who showed a reduction in serum CK levels and resolution of inflammatory signs in muscle magnetic resonance imaging (MRI) after tocilizumab treatment [54]. Another study revealed that tocilizumab treatment improved clinical and laboratory markers in a patient with an overlap syndrome comprising DM and systemic sclerosis who had been resistant to multiple therapies [55]. After continued treatment with tocilizumab, a patient with anti-Jo1- and Ro52-antibodies who suffered from recurring flares of myositis and arthritis with insufficient response to numerous medications showed clinical improvement and normalization of C-reactive protein and CK levels [56]. Tocilizumab did not fulfill the primary or secondary effectiveness outcomes in refractory DM and PM when studied in a randomized, double-blind, controlled phase II trial testing its efficacy in myositis patients [57].

4.5 Abatacept

Abatacept is a human fusion protein that inhibits T-cell costimulation by combining CTLA4 and the fragment-crystallizable region of IgG1. A recently published randomized, open-label, delayed-start trial in 20 individuals with refractory DM or PM indicated that abatacept therapy is effective [58]. The trial showed a significant improvement in muscle strength and health-related quality of life in half of the patients after treatment with i.v. abatacept for 6 months. The therapy was generally well tolerated. These positive results led to an ongoing phase III, randomized, double-blind trial evaluating the efficacy and safety of abatacept in myositis in which the primary endpoint was met in 56% of patients, but the p-value was denoted 0.08 [ClinicalTrials. gov identifier: NCT02971683].

4.6 Bimagrumab

The myostatin/activating type II receptor pathway controls muscle mass. In mouse studies, employing anti-ActRII antibodies causes muscular hypertrophy. The human anti-ActRII antibody is known as bimagrumab [59]. Bimagrumab was studied in 14 patients with IBM. Bimagrumab treatment resulted in an increase in muscle mass and body volume in the patients. In comparison to the placebo group, they improved their 6-minute walking distance [60]. However, in a recent double-blind multicenter trial, the primary endpoint (increasing muscle strength and 6-minute walking distance) was not, despite a favorable safety profile of the drug [61].

4.7 Sifalimumab

Overexpression of IFN-induced genes and IFN-regulated cytokines in blood samples from DM and PM suggests an essential involvement of interferon (IFN)/–mediated immunity in the pathogenesis of myositis [62, 63]. Sifalimumab is an anti-IFN monoclonal antibody whose effects in PM and DM were studied in a phase Ib randomized, double-blind, controlled clinical trial [64]. The suppression of the IFN signature in blood and muscle tissue in myositis patients treated with sifalimumab was linked to clinical improvement. Patients at baseline were identified as having IFN high vs. low gene expression profiles based on 13 type1 interferon-inducible genes.

Sifalimumab suppressed type I IFN expression by 66% in the blood and 47% in the muscle at day 98. Additionally, the levels of multiple dysregulated proteins (type 1 interferon-dependent and -independent) were measured in these patients and were found to be elevated in interferon high but not interferon low groups and correlated with MMT-8 scores. Patients with ≥15% MMT improvement showed greater neutralization of IFN signature than those with <15% improvement in both blood and muscle. Moreover, a reduced level of multiple T cell-associated proteins after sifalimumab but not placebo suggests a suppressive effect of blocking type I IFN signaling on T cell activation and chemoattraction that may lead to a reduction of T cell infiltration in the muscle of myositis patients.

4.8 JAK inhibitors

Ruxolitinib, a Janus kinase inhibitor, was recently shown to be successful in treating refractory dermatomyositis [65]. Ruxolitinib monotherapy led to rapid and significant improvement of dermatomyositis symptoms as the dermatomyositis was in remission by 12 months. Further case reports in juvenile dermatomyositis suggested the beneficial effect of JAK1/2 inhibitors, owing to primary role of constitutive type I IFN activation in the pathogenesis of the condition [66–68]. The use of another JAK inhibitor tofacitinib (a JAK 1/3 inhibitor) has been shown in a few case reports, comprising nine adult patients with refractory DM in total, with the majority improving clinically. Recently, preliminary results of an open-label pilot study evaluating tofacitinib in nine adult patients with refractory DM were presented. All nine patients showed minimal to moderate improvement after 12 weeks of treatment, with no reported serious adverse events. Further randomized controlled trials are expected to evaluate the efficacy and safety of JAK inhibitors.

4.9 Basiliximab

Basiliximab is an interleukin-2 receptor (IL-2R; CD25) chimeric monoclonal antibody that binds to IL-2 receptor on the activated T cells. The expression of interleukin – 2 receptor-α (IL-2Rα, or CD25) is especially upregulated on activated T and B cells. A small amount of IL-2Rα is also present in ordinary healthy people on inactive T and B cells and serum as soluble IL-2 receptor (sIL-2R). The increase in the expression of IL-2Rα, as well as sIL-2R, occurs in autoimmune diseases. One rationale for basiliximab use in myositis is that sIL-2R is correlated with disease activity in some DM/JDM patients.

Jing Zou et al. reported a case series of four adult amyopathic DM patients (positive anti-MDA5 antibody) who had failed conventional therapy. Three of four patients with rapidly progressing ILD demonstrated improved survival, reduction in ferritin levels, and improved lung functions with the use of two doses of 20 mg IV basiliximab 7 days apart. Subsequent trials are awaited.

4.10 Belimumab

Belimumab is a recombinant, fully human, monoclonal antibody directed against the cytokine BLyS, also known as B-cell activating factor (BAFF). It belongs to the tumor necrosis factor (TNF) superfamily and plays a central role in B-cell survival and function. A 40-week multicenter randomized, double-blind placebo-controlled trial with a 24-week open-label phase was conducted to assess the safety and efficacy of belimumab for IIM patients [69]. All patients met Peter and Bohan criteria and ACR 2017 classification criteria of polymyositis/dermatomyositis (PM/DM) with PM diagnosis adjudication. Refractory IIM was defined as inadequate response/intolerance to 3 months of glucocorticoids and/or at least one immunosuppressive agent (IS). Standard Core Set Measures (CSM) with MMT8 < 125/150 were used to define active disease. Patients on standard of care (SoC) therapy were randomized 1:1 to IV belimumab 10 mg/kg or placebo for 40 weeks followed by 24 weeks of belimumab 10 mg/kg in open-label phase. The study reported a numerically higher proportion of patients on belimumab reaching definition of improvement (DOI) vs. on standard of care (SoC) only arm. A higher proportion of patients on Belimumab achieved sustained moderate or major total improvement score (TIS) at 40 and 64 weeks compared with SoC. Detected differences were not statistically significant; however, the sample size was small.

	Trial	Results
Rituximab	1. **Rituximab in Myositis (RIM) trial**: 195 patients were randomized, double-blind, and placebo-controlled in this study (75 with PM,72 with DM, and 48 with JDM; all refractory to glucocorticoid therapy and at least one immunosuppressive drug) RIM trial-related research 2. **Efficacy of Rituximab in ASS trial**: Registry-based study of 43 patients evaluating the clinical response to many rituximab cycles in individuals with and without antisynthetase antibodies	83% of patients satisfied the definition of improvement. The steroid-sparing effect of rituximab was statistically significant. The most prevalent side effects of rituximab were infections. • Antisynthetase and anti-Mi-2 autoantibodies, as well as the juvenile DM subgroup and reduced disease damage, were all strong predictors of clinical improvement and rituximab response. • In adult DM and JDM patients, the addition of rituximab to conventional treatment resulted in significant improvements in cutaneous disease activity. Only the antibody-positive group demonstrated a substantial steroid-sparing effect, while both groups improved clinically independent of antibody status.
Anakinra	Case study based on 15 individuals with refractory myositis	Seven individuals had a clinical response to the International Myositis Assessment and Clinical Studies' core set of disease activity markers (IMACS), and functional index scores of four of them improved.
Infliximab	1. Randomized, double-blind, placebo-controlled trial investigating the efficacy of infliximab in refractory PM and DM. 2. Retrospective study of eight patients with refractory dermatomyositis or polymyositis. 3. Pilot study of 13 individuals with refractory IIM.	• Infliximab was effective in four out of twelve individuals. • However, myositis worsened in two people who appeared to respond to infliximab, and resuming the treatment was connected to anaphylaxis and the production of anti-dsDNA antibodies. The use of infliximab was associated with enhanced muscular strength and fatigue, but only a partial reduction in blood CK levels. Treatment with infliximab (four 5 mg/kg body weight infusions over 14 weeks) was found to be ineffective.

	Trial	Results
Tocilizumab	1. Case report of two patients with refractory Jo-1-positive PM 2. Case study of a patient with an overlap syndrome comprising DM and systemic sclerosis who had been resistant to multiple therapies. 3. Case study of a patient with anti-Jo1- and Ro52-antibodies who suffered from recurring flares of myositis and arthritis with insufficient response to numerous medications 4. Randomized, double-blind, controlled phase-II trial testing its efficacy in myositis patients.	Tocilizumab therapy resulted in a decrease in blood CK levels and the remission of inflammatory signals in muscle magnetic resonance imaging (MRI). Improved clinical and laboratory markers on treatment with tocilizumab. Following continued tocilizumab treatment, the patient's clinical condition improved, and his C-reactive protein and creatine kinase values returned to normal. In refractory DM and PM, tocilizumab did not meet the primary or secondary effectiveness outcomes.
Abatacept	Randomized, open-label, delayed-start trial in 20 individuals with refractory DM or PM	After 6 months of therapy with i.v. abatacept, half of the patients had a significant improvement in muscular strength and health-related quality of life.
Bimagrumab	1. Clinical trial in 14 patients with IBM 2. Double-blind multicenter trial	• Patients who received bimagrumab experienced an increase in muscle mass and body volume. • They improved their 6-minute walking distance when compared to the placebo group. Despite the drug's favorable safety profile, the primary goal (increased muscular strength and 6-minute walking distance) was not fulfilled.
Sifalimumab	Phase-Ib randomized, double-blind, controlled clinical trial.	• Clinical improvement was associated with the reduction of the IFN signature in blood and muscle tissue in myositis patients treated with sifalimumab. • Patients with $\geq$15% MMT improvement showed greater neutralization of IFN signature than those with <15% improvement in both blood and muscle. • A reduced level of multiple T cell-associated proteins after sifalimumab but not placebo, suggests a suppressive effect of blocking type I IFN signaling on T cell activation and chemoattraction that may lead to a reduction of T cell infiltration in the muscle of myositis patients
Ruxolitinib	Case reports (nine adult patients with refractory DM)	• Majority improved clinically with Ruxolitinib treatment. • After 12 weeks of therapy, all nine patients demonstrated modest to moderate improvement, with no significant side effects noted.

	Trial	Results
Basiliximab	Jing Zou et al. case series of four adult amyopathic DM patients (positive anti-MDA5 antibody) who had failed conventional therapy.	With two doses of 20 mg IV basiliximab given 7 days apart, three of four patients with rapidly progressing ILD showed better survival, reduced ferritin levels, and improved lung function.
Belimumab	40-week multi-center randomized, double-blind placebo-controlled trial with 24 weeks open-label phase in IIM patients.	• Higher proportion of patients on belimumab reaching definition of improvement (DOI) vs. on standard of care (SoC) only arm. • When compared to SOC, a larger percentage of patients on Belimumab had a sustained moderate or substantial total improvement score (TIS) at 40 and 64 weeks.

5. Conclusion

The introduction of biologic therapies has held considerable promise for autoimmune diseases, allowing us to translate our knowledge of specific disease pathophysiology processes into treatments that target certain autoimmune disorders. One challenge in examining prospective biologic and other treatments for IIM in the future is the scarcity of information. There is mounting evidence that biologic therapy in IIM can help patients with refractory disease by improving muscle strength, lowering biochemical markers of muscle inflammation, and weaning them off of glucocorticoids. The fact that IIMs are rare diseases poses a substantial hurdle, as it limits the number of people who can participate in clinical studies. Furthermore, while substantial progress has been made in understanding the etiology of IIM, much remains unknown about the immunological systems that underpin these disorders. Rituximab has been the most thoroughly studied of the biologic medicines, and it appears to be successful in people with PM, DM, and JDM. Other agents are constrained by a lack of trial data and small sample sizes, notwithstanding their promise. Because of the significant basic research evidence that interferon is essential to the etiopathogenesis of IIM disease, sifalimumab must currently be considered the biologic with the most potential in the future. The biologic rationales for the medicines may be discussed in-depth, and the enthusiasm for the future is genuine. However, most of these treatments are likely to be 3–5 years, if not more, away from the clinician's repertoire and patient therapy. It's anyone's guess what the future holds and how much the drugs will cost.

Conflict of interest

The authors declare no conflict of interest.

Author details

Ashwin Parchani, Aditya Sudan, Shiana Singh, Arpit Singh and Monika Pathania*
All India Institute of Medical Sciences, Rishikesh, India

*Address all correspondence to: anshupathania27@gmail.com

References

[1] Schett G, McInnes IB, Neurath MF. Reframing immune-mediated inflammatory diseases through signature cytokine hubs. The New England Journal of Medicine. 2021;**385**(7)

[2] Slevin SM, Egan LJ. New insights into the mechanisms of action of anti-tumor necrosis factor-α monoclonal antibodies in inflammatory bowel disease. Inflammatory Bowel Diseases. 2015;**21**(12)

[3] Pircher J, Merkle M, Wörnle M, Ribeiro A, Czermak T, Stampnik Y, et al. Prothrombotic effects of tumor necrosis factor alpha in vivo are amplified by the absence of TNF-alpha receptor subtype 1 and require TNF-alpha receptor subtype 2. Arthritis Research & Therapy. 2012;**14**(5)

[4] Qu Y, Zhao G, Li H. Forward and reverse signaling mediated by transmembrane tumor necrosis factor-alpha and TNF receptor 2: Potential roles in an immunosuppressive tumor microenvironment. Frontiers in Immunology. 2017;**8**

[5] Lima MSR, de Lima VC, Piuvezam G, de Azevedo KP, Maciel BLL, Morais AHDA. Mechanisms of action of molecules with anti-TNF-alpha activity on intestinal barrier inflammation: A systematic review protocol. Medicine (United States). 2019;**98**

[6] Vitale A, Obici L, Cattalini M, Lopalco G, Merlini G, Ricco N, et al. Biotechnological agents for patients with tumor necrosis factor receptor associated periodic syndrome—Therapeutic outcome and predictors of response: Real-life data from the AIDA network. Frontiers in Medicine. 2021;**8**

[7] Schett G. Physiological effects of modulating the interleukin-6 axis. Rheumatology (United Kingdom). 2018;**57**

[8] Favalli EG. Understanding the role of Interleukin-6 (IL-6) in the joint and beyond: A comprehensive review of IL-6 inhibition for the Management of Rheumatoid Arthritis. Rheumatology and Therapy. 2020;**7**

[9] da Gonçalves-de-Albuquerque S, Pessoa-e-Silva R, Trajano-Silva LA, de Goes TC, de Morais RC, Oliveira CN, et al. The equivocal role of Th17 cells and neutrophils on immunopathogenesis of leishmaniasis. Frontiers in Immunology. 2017;**8**

[10] Domeier PP, Rahman ZSM. Regulation of B cell responses in sle by three classes of interferons. International Journal of Molecular Sciences. 2021;**22**

[11] Absher DM, Li X, Waite LL, Gibson A, Roberts K, Edberg J, et al. Genome-wide DNA methylation analysis of systemic lupus erythematosus reveals persistent hypomethylation of interferon genes and compositional changes to CD4+ T-cell populations. PLoS Genetics. 2013;**9**(8)

[12] Bridgewood C, Sharif K, Sherlock J, Watad A, McGonagle D. Interleukin-23 pathway at the enthesis: The emerging story of enthesitis in spondyloarthropathy. Immunological Reviews. 2020;**294**

[13] BabaieF,HasankhaniM,MohammadiH, Safarzadeh E, Rezaiemanesh A, Salimi R, et al. The role of gut microbiota and IL-23/IL-17 pathway in ankylosing spondylitis immunopathogenesis: New insights and updates. Immunology Letters. 2018;**196**

[14] Liu T, Li S, Ying S, Tang S, Ding Y, Li Y, et al. The IL-23/IL-17 pathway in inflammatory skin diseases: From bench to bedside. Frontiers in Immunology. 2020;**11**

[15] Puig L, Lebwohl M, Bachelez H, Sobell J, Jacobson AA. Long-term efficacy and safety of brodalumab in the treatment of psoriasis: 120-week results from the randomized, double-blind, placebo- and active comparator–controlled phase 3 AMAGINE-2 trial. Journal of the American Academy of Dermatology. 2020;**82**(2)

[16] Herrero-Beaumont G, Martínez Calatrava MJ, Castañeda S. Abatacept mechanism of action: Concordance with its clinical profile. Reumatología Clínica (English Edition). 2012;**8**(2)

[17] Cagnotto G, Willim M, Nilsson JÅ, Compagno M, Jacobsson LTH, Saevarsdottir S, et al. Abatacept in rheumatoid arthritis: Survival on drug, clinical outcomes, and their predictors - data from a large national quality register. Arthritis Research & Therapy. 2020;**22**(1)

[18] Humbel M, Bellanger F, Fluder N, Horisberger A, Suffiotti M, Fenwick C, et al. Restoration of NK cell cytotoxic function with Elotuzumab and daratumumab promotes elimination of circulating plasma cells in patients with SLE. Frontiers in Immunology. 2021;**12**

[19] Mould DR. The pharmacokinetics of biologics: A primer. Digestive Diseases. 2015;**33**(suppl. 1):61-69

[20] Lobo ED, Hansen RJ, Balthasar JP. Antibody pharmacokinetics and pharmacodynamics. Journal of Pharmaceutical Science. 2004;**93**(11): 2645-2668

[21] Datta-Mannan A. Mechanisms influencing the pharmacokinetics and disposition of monoclonal antibodies and peptides. Drug Metabolism and Disposition. 2019;**47**(10):1100-1110

[22] Thomas VA, Balthasar JP. Understanding inter-individual variability in monoclonal antibody disposition. Antibodies. 2019;**8**:56

[23] Posner J, Barrington P, Brier T, Datta-Mannan A. Monoclonal antibodies: Past, present and future. In: Barrett JE, Page CP, Michel MC, editors. Concepts and Principles of Pharmacology: 100 Years of the Handbook of Experimental Pharmacology. Cham, Switzerland: Springer International Publishing; 2019. pp. 81-141

[24] Wright A, Sato Y, Okada T, Chang K, Endo T, Morrison S. In vivo trafficking and catabolism of IgG1 antibodies with Fc associated carbohydrates of differing structure. Glycobiology. 2000;**10**:1347-1355

[25] Israel EJ, Wilsker DF, Hayes KC, Schoenfeld D, Simister NE. Increased clearance of IgG in mice that lack beta 2-microglobulin: Possible protective role of FcRn. Immunology. 1996;**89**:573-578

[26] Unghans RP, Anderson CL. The protection receptor for IgG catabolism is the beta2-microglobulin-containing neonatal intestinal transport receptor. Proceedings of the National Academy Science USA. 1996;**93**:5512-5516

[27] Liu L. Pharmacokinetics of monoclonal antibodies and fc-fusion proteins. Protein & Cell. 2018;**9**:15-32

[28] Zhou SY, Shu C, Korth-Bradley J, Raible D, Palmisano M, Wadjula J, et al. Integrated population pharmacokinetics of etanercept in healthy subjects and in patients with rheumatoid arthritis ankylosing spondylitis. The Journal of Clinical Pharmacology. 2011;**51**(6):864-875

[29] Yim DS, Zhou H, Buckwalter M, Nestorov I, Peck CC, Lee H. Population pharmacokinetic analysis and simulation of the time-concentration profile of etanercept in pediatric patients with juvenile rheumatoid arthritis. Journal of Clinical Pharmacology. 2005;**45**(3):246-256. DOI: 10.1177/0091270004271945. (PubMed ID 15703360)

[30] FDA Approved Drug Products: Enbrel (etanercept) injection solution

[31] Klotz U, Teml A, Schwab M. Clinical pharmacokinetics and use of infliximab. Clinical Pharmacokinetics. 2007;**46**(8):645-660

[32] HUMIRA (adalimumab), Abbott Laboratories - package insert. 2022. Accessdata.fda.gov. [cited 26 May 2022]

[33] Kivitz A, Segurado OG. HUMIRA pen: A novel autoinjection device for subcutaneous injection of the fully human monoclonal antibody adalimumab. Expert Review of Medical Devices. 2007;**4**(2):109-116. DOI: 10.1586/17434440.4.2.109

[34] Scheinfeld N. Adalimumab (HUMIRA): a review. Journal of Drugs in Dermatology. 2003;**2**(4):375-377

[35] FDA Approved Drug Products: KINERET (anakinra) subcutaneous injection

[36] FDA Clinical Pharmacology and Biopharmaceutics Review: Kineret (anakinra) subcutaneous injection

[37] FDA Approved Drug Products: ACTEMRA (tocilizumab) injection, for intravenous or subcutaneous use

[38] Abdallah H, Hsu JC, Lu P, Fettner S, Zhang X, Douglass W, et al. Pharmacokinetic and pharmacodynamic analysis of subcutaneous tocilizumab in patients with rheumatoid arthritis from 2 Randomized, Controlled Trials: SUMMACTA and BREVACTA. Journal of Clinical Pharmacology. 2017;**57**(4):459-468

[39] Ramiro S, Smolen JS, Landewé R, et al. Pharmacological treatment of psoriatic arthritis: A systematic literature review for the 2015 update of the EULAR recommendations for the management of psoriatic arthritis. Annals of the Rheumatic Diseases. 2016;**75**:490

[40] Fasano S, Gordon P, Hajji R, Loyo E, Isenberg DA. Rituximab in the treatment of inflammatory myopathies: A review. Rheumatology (Oxford, England). 2017;**56**(1):26-36

[41] Oddis CV, Reed AM, Aggarwal R, Rider LG, Ascherman DP, Levesque MC, et al. Rituximab in the treatment of refractory adult and juvenile dermatomyositis and adult polymyositis: A randomized, placebo-phase trial. Arthritis and Rheumatism. 2013;**65**(2):314-324

[42] Aggarwal R, Bandos A, Reed AM, Ascherman DP, Barohn RJ, Feldman BM, et al. Predictors of clinical improvement in rituximab-treated refractory adult and juvenile dermatomyositis and adult polymyositis. Arthritis Rheumatology Hoboken NJ. 2014;**66**(3):740-749

[43] Aggarwal R, Oddis CV, Goudeau D, Koontz D, Qi Z, Reed AM, et al. Autoantibody levels in myositis patients correlate with clinical response during B cell depletion with rituximab. Rheumatology (Oxford, England). 2016;**55**(6):991-999

[44] Leclair V, Galindo-Feria AS, Dastmalchi M, Holmqvist M, Lundberg IE. Efficacy and safety of rituximab in anti-synthetase antibody positive and negative subjects with idiopathic

inflammatory myopathy: A registry-based study. Rheumatology (Oxford, England). 2019;**58**(7):1214-1220

[45] de Souza FHC, Miossi R, de Moraes JCB, Bonfá E, Shinjo SK. Favorable rituximab response in patients with refractory idiopathic inflammatory myopathies. Advanced in Rheumatology. 2018;**58**(1):31

[46] Zong M, Dorph C, Dastmalchi M, Alexanderson H, Pieper J, Amoudruz P, et al. Anakinra treatment in patients with refractory inflammatory myopathies and possible predictive response biomarkers: A mechanistic study with 12 months follow-up. Annals of the Rheumatic Diseases. 2014;**73**(5):913-920

[47] Oddis CV, Aggarwal R. Treatment in myositis. Nature Reviews Rheumatology. 2018;**14**(5):279-289

[48] Schiffenbauer A, Garg M, Castro C, Pokrovnichka A, Joe G, Shrader J, et al. A randomized, double-blind, placebo-controlled trial of infliximab in refractory polymyositis and dermatomyositis. Seminars in Arthritis and Rheumatism. 2018;**47**(6):858-864

[49] Hengstman GJD, van den Hoogen FHJ, van Engelen BGM. Treatment of dermatomyositis and polymyositis with anti-tumor necrosis factor-alpha: Long-term follow-up. European Neurology. 2004;**52**(1):61-63

[50] Efthimiou P, Schwartzman S, Kagen LJ. Possible role for tumour necrosis factor inhibitors in the treatment of resistant dermatomyositis and polymyositis: A retrospective study of eight patients. Annals of the Rheumatic Diseases. 2006;**65**(9):1233-1236

[51] Dastmalchi M, Grundtman C, Alexanderson H, Mavragani CP, Einarsdottir H, Helmers SB, et al. A high incidence of disease flares in an open pilot study of infliximab in patients with refractory inflammatory myopathies. Annals of the Rheumatic Diseases. 2008;**67**(12):1670-1677

[52] Hengstman GJD, De Bleecker JL, Feist E, Vissing J, Denton CP, Manoussakis MN, et al. Open-label trial of anti-TNF-alpha in dermato- and polymyositis treated concomitantly with methotrexate. European Neurology. 2008;**59**(3-4):159-163

[53] Ishikawa Y, Yukawa N, Ohmura K, Hosono Y, Imura Y, Kawabata D, et al. Etanercept-induced anti-Jo-1-antibody-positive polymyositis in a patient with rheumatoid arthritis: A case report and review of the literature. Clinical Rheumatology. 2010;**29**(5):563-566

[54] Narazaki M, Hagihara K, Shima Y, Ogata A, Kishimoto T, Tanaka T. Therapeutic effect of tocilizumab on two patients with polymyositis. Rheumatology (Oxford, England). 2011;**50**(7):1344-1346

[55] Kondo M, Murakawa Y, Matsumura T, Matsumoto O, Taira M, Moriyama M, et al. A case of overlap syndrome successfully treated with tocilizumab: A hopeful treatment strategy for refractory dermatomyositis? Rheumatology (Oxford, England). 2014;**53**(10):1907-1908

[56] Murphy SM, Lilleker JB, Helliwell P, Chinoy H. The successful use of tocilizumab as third-line biologic therapy in a case of refractory anti-synthetase syndrome. Rheumatology (Oxford, England). 2016;**55**(12):2277-2278

[57] Abstract Supplement ACR. Convergence 2020. Arthritis & Rhematology. 2020;**72**(S10):1-4259

[58] Tjärnlund A, Tang Q, Wick C, Dastmalchi M, Mann H, Tomasová

Studýnková J, et al. Abatacept in the treatment of adult dermatomyositis and polymyositis: A randomised, phase IIb treatment delayed-start trial. Annals of the Rheumatic Diseases. 2018;**77**(1):55-62

[59] Lach-Trifilieff E, Minetti GC, Sheppard K, Ibebunjo C, Feige JN, Hartmann S, et al. An antibody blocking activin type II receptors induces strong skeletal muscle hypertrophy and protects from atrophy. Molecular and Cellular Biology. 2014;**34**(4):606-618

[60] Amato AA, Sivakumar K, Goyal N, David WS, Salajegheh M, Praestgaard J, et al. Treatment of sporadic inclusion body myositis with bimagrumab. Neurology. 2014;**83**(24):2239-2246

[61] Ioannis M, Foivos P, Dimitrios K. A review on the treatment of sporadic inclusion body myositis with Bimagrumab and alemtuzumab. The International Journal of Neuroscience. 2019;**129**(3):297-302

[62] Walsh RJ, Kong SW, Yao Y, Jallal B, Kiener PA, Pinkus JL, et al. Type I interferon-inducible gene expression in blood is present and reflects disease activity in dermatomyositis and polymyositis. Arthritis and Rheumatism. 2007;**56**(11):3784-3792

[63] Greenberg SA, Pinkus JL, Pinkus GS, Burleson T, Sanoudou D, Tawil R, et al. Interferon-alpha/beta-mediated innate immune mechanisms in dermatomyositis. Annals of Neurology. 2005;**57**(5):664-678

[64] Higgs BW, Zhu W, Morehouse C, White WI, Brohawn P, Guo X, et al. A phase 1b clinical trial evaluating sifalimumab, an anti-IFN-α monoclonal antibody, shows target neutralisation of a type I IFN signature in blood of dermatomyositis and polymyositis patients. Annals of the Rheumatic Diseases. 2014;**73**(1):256-262

[65] Hornung T, Janzen V, Heidgen FJ, Wolf D, Bieber T, Wenzel J. Remission of recalcitrant dermatomyositis treated with ruxolitinib. The New England Journal of Medicine. 2014;**371**(26):2537-2538

[66] Sabbagh S, Almeida de Jesus A, Hwang S, Kuehn HS, Kim H, Jung L, et al. Treatment of anti-MDA5 autoantibody-positive juvenile dermatomyositis using tofacitinib. Brain. Journal of Neurology. 2019;**142**(11):e59

[67] Papadopoulou C, Hong Y, Omoyinmi E, Brogan PA, Eleftheriou D. Janus kinase 1/2 inhibition with baricitinib in the treatment of juvenile dermatomyositis. Brain: A Journal of Neurology. 2019;**142**(3):e8

[68] Heinen A, Schnabel A, Brück N, Smitka M, Wolf C, Lucas N, et al. Interferon signature guiding therapeutic decision making: Ruxolitinib as first-line therapy for severe juvenile dermatomyositis? Rheumatology. 2021;**60**(4):e136-e138

[69] Galina Marder. Safety and Efficacy of Belimumab in the Treatment of Adult Idiopathic Inflammatory Myositis (Polymyositis and Dermatomyositis). Available from: https://acrabstracts.org/tag/myopathies/